THE EMERGENCE OF PROBABILITY

THE EMERGENCE OF PROBABILITY

A PHILOSOPHICAL STUDY OF
EARLY IDEAS ABOUT PROBABILITY, INDUCTION
AND STATISTICAL INFERENCE

IAN HACKING

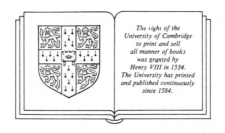

The right of the
University of Cambridge
to print and sell
all manner of books
was granted by
Henry VIII in 1534.
The University has printed
and published continuously
since 1584.

CAMBRIDGE UNIVERSITY PRESS

CAMBRIDGE
LONDON NEW YORK NEW ROCHELLE
MELBOURNE SYDNEY

Published by the Press Syndicate of the University of Cambridge
The Pitt Building, Trumpington Street, Cambridge CB2 1RP
32 East 57th Street, New York, NY 10022, USA
296 Beaconsfield Parade, Middle Park, Melbourne 3206, Australia

First published 1975
Reprinted 1978
First paperback edition 1984

Printed in Great Britain
by J. W. Arrowsmith Ltd.,
Winterstoke Road, Bristol BS3 2NT

Library of Congress catalogue card number: 74–82224

ISBN 0 521 20460 7 hard covers
ISBN 0 521 31803 3 paperback

FOR DANIEL

CONTENTS

probability on propositions, namely made them worthy of approval. But it did so in virtue of the frequency with which it made correct predictions. This transformation from sign into evidence is the key to the emergence of a concept of probability that is dual in the sense of Chapter 2.

Contents

1

AN ABSENT FAMILY OF IDEAS

In 1865 Isaac Todhunter published *A History of the Mathematical Theory of Probability from the Time of Pascal to that of Laplace*. It remains an authoritative survey of nearly all work between 1654 and 1812. Its title is exactly right. There was hardly any history to record before Pascal, while after Laplace probability was so well understood that a page-by-page account of published work on the subject became almost impossible. Just six of the 618 pages of text in Todhunter's book discuss Pascal's predecessors. Subsequent scholarship can do better but even now we can light on only a few earlier memoranda and unpublished notes. Yet in 'the time of Pascal' all manner of citizens recognized the emergent idea of probability. A philosophical history must not only record what happened around 1660, but must also speculate on how such a fundamental concept as probability could emerge so suddenly.

Probability has two aspects. It is connected with the degree of belief warranted by evidence, and it is connected with the tendency, displayed by some chance devices, to produce stable relative frequencies. Neither of these aspects was self-consciously and deliberately apprehended by any substantial body of thinkers before the time of Pascal. There have been several tentative explanations of this fact. I shall describe them briefly here, but we shall see that none of them is at all satisfactory.

First let us run through some quite well known facts and conjectures about the prehistory of randomness. Since gambling is ancient and possibly primeval we expect early ideas of probability. In her engaging book on the history of this subject, F. N. David [1962] speculates that gambling may be a first invention of human society. Her clue to this is the *talus*. This most common randomizer of ancient times, is a predecessor of the die: the astragalus or talus is the 'knucklebone' or heel bone of a running animal. In creatures such as deer, horse, oxen, sheep and hartebeeste this bone is so formed that when it is thrown to land on a level surface it can come to rest in only four ways. Well polished and often engraved examples are regularly found on the sites of ancient Egypt. Tomb illustrations and scoring boards make it virtually certain that these were used for gaming.

Similar sheep-like tali begin to occur on Sumerian and Assyrian sites. Well polished, oft-used knucklebones are found even in Paleolithic dwellings. But, unlike the Sumerian sheep bones, these are misleading. A casual glance at a photograph makes them seem just like Egyptian tali, but in real life they are enormous, the heel bones of oxen or larger beasts. The more slender part of these large bones is an unmistakeable handle, and the blunt end a natural hammer head. We have no need to imagine these are the mighty dice of debauched and giant cavemen. They are tools. We have no evidence of possible gambling devices before the Sumerian and Assyrian sites, and no proof before Egypt. That is old enough.

It seems to follow that empirical frequencies and averages should be as old as the rolling of such ancient bones. Nor were tali the only randomizers. Deciding by lot is familiar from the Talmud, from which A. M. Hasofer [1967] has recently dug a little probability arithmetic. Nor can we suppose that gambling was the invention of a single people who passed it on to their cultural heirs. In one of the very first books of probability theory, Pierre Rémond de Montmort [1708, p. xii] works out the probability laws of some pastimes that Jesuit missionaries found current among seventeenth century Huron. And so on: it is hard to find a place where people use no randomizers. Yet theories of frequency, betting, randomness and probability appear only recently. No one knows why. We can canvass a few of the proposed answers.

First, it has been urged that an obsession with determinism precluded any thought about randomness. But this is doubly absurd. It is absurd because it is anachronistic: the model for determinism that it invokes is the recent one that became dominant in the seventeenth century, and it makes no attempt to understand earlier concepts of freedom and the rule of natural law. This first absurdity points at once to the second one. Europe began to understand concepts of randomness, probability, chance and expectation precisely at that point in its history when theological views of divine foreknowledge were being reinforced by the amazing success of mechanistic models. A good many different kinds of determinism have appeared in various ages and cultures. Most of us think only of the mechanistic attitude to causation that first came into being in the seventeenth century. Far from this 'mechanical' determinism precluding an investigation of chance, it was its accompaniment. A paradoxical but better conjecture would be that this specific mode of determinism is essential to the formation of concepts of chance and probability. We cannot appeal to some ancient deterministic fetish to explain lack of probability: we would do better to explain lack of probability by lack of such a fetish.

Second, and perhaps more satisfactory, is a shift of focus from determinism of our modern kind back to a notion of gods settling what will happen. Cutting up fowl to predict the future is, if done honestly and with as little interpretation as possible, a kind of randomization. But chicken guts are hard to read and invite flights of fancy or corruption. We know that the Israelites, ever sceptical of their conniving priests, preferred the lot whose meaning is open for all to read. Lotteries and dice make a good way of consulting gods directly. But then (so it has been argued) it would be naughty or even impious to try to compute what the gods will say. The role of dicing in divination might preclude critical investigation of the laws of randomness.

This explanation will not do. There were plenty of impious people gambling like mad. Marcus Aurelius was so obsessed with throwing dice to pass the time that he was regularly accompanied by his personal croupier. Less reputable gentlemen are also well documented. Someone with only the most modest knowledge of probability mathematics could have won himself the whole of Gaul in a week. The fact that some people were pious and others superstitious, far from preventing the opportunists of an opulent empire discovering some elementary arithmetic of dice, is a positive incentive.

Next it has been suggested that in order to conceive of probability laws we need some easily understood empirical examples. Even today the best of probability textbooks instill mathematical intuitions by example rather than precept. The first examples always employ what Jerzy Neyman [1950, p. 15] called a *Fundamental Probability Set* of equally probable alternatives. Only after the student grasps this idea does he progress to sets whose alternatives are not equiprobable. So it is suggested that in ancient times

3

equiprobable sets had not been noticed. In particular, tali do not provide us with a set of four equal chances. Moreover, the distribution of chances varies from talus to talus according to the distribution of mass in the heelbone. Subsequent dice may also have been imperfect. There did not exist an *F.P.S.* to give us the idea of probability.

This explanation is defective in point of fact. Dice of ivory and other uniform materials were made long ago. The dice in the cabinets of the Cairo Museum of Antiquities, which the guards kindly allowed me to roll for a long afternoon, appear to be exquisitely well balanced. Indeed a couple of rather irregular looking ones were so well balanced as to suggest that they had been filed off at this or that corner just to make them equiprobable. There were plenty of stochastically sound Fundamental Probability Sets to be had in ancient times.

Next, there is the economic theory, for which some evidence is presented by L. E. Maistrov [1974]. The basic doctrine is that a science develops to answer economic needs. It would be glib to protest that those wretched gamblers of times gone by had plenty of economic needs but did not invent a calculus. For the economic needs referred to are not those of whimsical overlords but the means of production and the organization of the state. An undogmatic version of this doctrine must be right. There are two ways in which a science develops: in response to problems which it itself creates, and in response to problems that are forced on it from the outside. Only very recently has probability theory been hardy enough to create its own problems and generate its own programmes of research. The stimulus used to come from other disciplines. In the seventeenth century insurance and annuities were a focus of attention. In the eighteenth, a theory of measurement was needed, chiefly but not solely for astronomy. In the latter part of the nineteenth century analysis of biological data demanded a mathematics that created 'biometrics'. Statistical mechanics required a deeper analysis of probability concepts. The needs of agricultural and medical experiments produced the bulk of the really great statistical theory in Western Europe in the early part of the twentieth century, while a quite different sort of problem led to the new Russian theories on measure-theoretic foundations of probability. Each of these needs, with the possible exception of the theory of astronomical measurement whose culmination is Laplace and Gauss (no mere anomaly!) can, it is clear, be cast as an economic need. So an economic history of probability will fare quite well. It cheerfully cites Huygens, De Moivre, Gauss, Galton, K. Pearson, Fisher, A. Markov, Von Mises, Kolmogorov, Neyman, Wald and Savage as workers whose problem-situation originates in the means of production of society.

Despite the superficial success of such a selective story, I do not believe that any economic theory about the *origin* of probability can carry conviction. It is true that we first find European calculations on chances in work like that of Pacioli [1494], a book famous as the origin of double-entry book-keeping. But what is notable is not that problems on chance occur in early works of arithmetic chiefly aimed at the new commerce, but that these books were quite unable to solve the problems. No one could solve them until about 1660, and then everyone could. Or to take annuities, it is true that they were used to finance the Netherlands, and that at a time when Holland was the asylum of Europe, annuity mathematics was invented by politically influential students of Descartes. But it was not this method of finance that made probability possible. We know from reading the third century Roman jurist Ullpian that, like Holland, his state also derived cash by selling annuities [cf. Greenwood 1940]. Ullpian's annuities, although not based on any explicit actuarial or probabilistic reasoning, were well and perhaps even cunningly devised to serve the needs of the state and not its citizens. The Dutch, in contrast, calculated their annuities so badly that the towns were regularly losing money. In England, Isaac Newton himself could allow his imprimatur to appear on a book by Mabbut [1686] that did not allow for the age of annuitants in computing their annual returns. This ignorant practice continued for over a century, despite the fact that the mathematicians constantly inveighed against it. In short, some skill at getting sensible annuities was common fifteen centuries ago and there was no theory, yet our greatest mathematician, who had substantial understanding of probability and whose contemporaries invented it, allowed non-actuarial annuity rates. Economic need does not, in this case, seem well correlated with growth of understanding; nor does growth of understanding seem well correlated with satisfaction of economic needs.

The economic account of the genesis of probability is external: it says that understanding is in response to a problem, usually an economic one. There is an alternative story which is internal. Mathematics, it is suggested, was not sufficiently rich in ideas to generate a probability calculus. Once probability did get going this opinion is right. Probability mathematics becomes serious just when limit theorems became possible; it could not be deeply serious before then. Yet the concept of probability does not appear to require limit theorems and a profiteer can do well without knowing anything much besides fairly simple arithmetic. Why did the arithmetic occur to no one? The ingenuous answer is that there was no arithmetic.

Even today some arithmetical juggling is needed for a beginner to get the feel of probabilistic calculations. The calculations use numbers. If a man has 37 matches in the box in his left pocket and 21 in the box in his right pocket and grabs either box at random, what is the probability that he gets to a point where he has one match in each pocket? These are calculations with numbers; when you can do them, you know quite a lot about probability. Now these calculations require some facility with figures. The Greeks, giving pride of place to geometry, lacked a perspicuous notation for numerals. So did their heirs. Perhaps a symbolism that makes addition and multiplication easy is a prerequisite for any rich concept of probability. Two pieces of circumstantial evidence may seem to support this view. First, probability mathematics is, like our system of numerals, almost certainly of Arabic origin. Indeed the old word for chance, namely 'hazard', is as arabic as 'algebra'. The first European probabilists were Italian, solving North African problems in hazard at the same time as they were advancing algebra.

There is a second item of evidence which each reader will judge as he will. In Europe we find glimmerings of a science of dicing in the fifteenth century. There is still not the slightest intimation that this might have applications to anything of interest to non-gamblers. The first inkling of that occurs in the middle of the seventeenth century. Yet there is a shred of evidence that there was a science of dicing millenia earlier, and, incomparably more important, that people knew how this science affected matters such as survey-sampling, which we would otherwise have thought of as a nineteenth century enterprise.

V. P. Godambe has recently drawn attention to a passage in the great Indian epic *Mahábarata*. It is very ancient but, according to scholars such as Renou and Filliozat [1947, p. 401] the present version was finished about A.D. 400. In the third book is an interpolation, the story of Nala. Kali, a demigod of dicing, has his eye on a delicious princess and is dismayed when Nala wins her hand. In revenge Kali takes possession of the body and soul of Nala, who in a sudden frenzy of gambling loses his kingdom. He wanders demented for many years. But then, following the advice of a snake-king whom he meets in the forest, he takes a job as charioteer to the foreign potentate Rtuparna. On a journey the latter flaunts his mathematical skill by estimating the number of leaves and of fruit on two great branches of a spreading tree. Apparently he does this on the basis of a single twig that he examines. There are, he avers, 2095 fruit. Nala counts all night and is duly amazed by the accuracy of this guess. Rtuparna, so often right in matters like this, accepts his due. In the translation of H. H. Milman [1860, p. 76] he says:

> I of dice possess the science
> and in numbers thus am skilled.

He agrees to teach this science to Nala in exchange for some lessons in horsemanship. At the end of this equestrian course in survey sampling Nala vomits out the poison of Kali and, restored to his usual form, wins back his kingdom in a fierce game in which he has to stake his ever-faithful bride.

Before learning the science of dice the bewitched Nala was an obsessive gambler, but after mastering the science he is able to place bets so he can recoup his birthright. That is evidence that in India, long ago, it was recognized that there was a genuine science to master, whereas in Europe this knowledge seems to have been lacking. More striking is the recognition that dicing has something to do with estimating the number of leaves on a tree. That indicates a very high level of sophistication. Even after the European invention of probability around 1660 it took some time before any substantial body of people could comprehend that decisive connection. Indeed, although the Nala story was almost the first piece of Sanskrit writing to be widely circulated in modern Europe and was much admired by the German romantics, no one paid any attention (so far as I know) to this curious insight about the connection between dicing and sampling.

On the other hand stories about dicing, and the loss of fortune thereby, occur frequently in Indian literature. One of these is the chief topic of Book II of the Mahábarata. A summary may be read in the survey of Indian literature written by M. Winternitz [1927, pp. 341–6]. To judge by Winternitz's survey, the dicing stories that recur throughout the whole literature have a predominantly moral overtone, and are intended to warn people against gaming. Yet the Nala story seems a positive incentive for mastering the science of dice and thereby reaping great gains when playing with less well-informed people. When the story is taken up in puritanical Jain literature, however, that profiteering element is completely ignored.

Indian mathematical texts ought to yield a rich reward to the student of probability. They have not yet been investigated with this end in view and it is unclear what will turn up. Take for example the mathematician Mahāvirācārya, whom his translator M. Rangācārya [1912, p. x] dates about the end of the ninth century A.D. Here we find a use of what modern probabilists call a 'Dutch book'. That is, a merchant 'secretly' bets with two different agents at discrepant odds, in such a way that no matter what actually happens, the merchant is guaranteed a profit [*ibid.*, pp. 162–3].

It is reasonable to guess, then, that a good deal of Indian probability lore is at present unknown to us. That accords well with the conjectures I have been describing in such a lukewarm way. Why was there no probability theory in the West before Pascal? Answers: a necessitarian view of the world, piety, lack of a place system of numeration and of economic incentive. Corollary: impiety, arithmetic, plenty of trading and different concepts of causality should be conducive to the formation of probability mathematics. Confirmation: two millenia ago India had an advanced merchant system, it had handy numerals, and both its piety and its theories of causation were not at all in the European mould. In that society we find hints of a hitherto unknown theory of probability.

Unfortunately no explanation, of the sort I have been describing, can explain very much. As the Indian example reminds us, these explanations although instructive are pretty nebulous. Perhaps the explanations and conjectures are directed at the wrong question. All take for granted that there existed an intellectual object – a concept of probability – which was not adequately thought about nor sufficiently subject to mathematical reflection. So one asks, what technology was missing? What incentive was absent? These questions are appropriate only if the conceptual scheme of those

8

earlier times had within it a concept of probability. If there was no such object, then all the questions are idle.

Men did make randomizers, and did generate stable frequencies with dice. They did draw inductive inferences whose conclusions were merely probable. It in no way follows that they had anything like our conception of these things. We should not ask, why did people fail to study these objects? We should ask instead, how did these objects of thought come into being?

All the conjectural explanations I have described try to locate something lacking in pre-Pascalian times. No one denies that arithmetic and nascent capitalism were lacking, nor that one or the other may be essential to the development of probability theory, once probability is a possible object of thought. We should, however, try to find out how probability became possible at all.

We do not ask how *some* concept of probability became possible. Rather we need to understand a quite specific event that occurred around 1660: the emergence of *our* concept of probability. If there were Indian concepts of probability 2000 years ago, they doubtless arose from a transformation quite different from the one we witness in European history. From a purely historical point of view, both transformations may be of equal interest. But for me the search for preconditions is more than an attempt at historical explanation. I am inclined to think that the preconditions for the emergence of our concept of probability determined the very nature of this intellectual object, 'probability', that we still recognize and employ and which, as philosophers, we still argue about. The preconditions for the emergence of probability determined the space of possible theories about probability. That means that they determined, in part, the space of possible interpretations of quantum mechanics, of statistical inference, and of inductive logic.

I cannot here establish a claim so grand, so pretentious. It is a motivation for research, not a proven conclusion. It indicates only a direction in which to proceed. It makes the prehistory of probability more important than the history. But prehistory must be properly understood. We are looking for neither precursors nor anticipations of our ideas. The preconditions for probability will consist in something that is not probability but which was, through something like a mutation, transformed into probability. This non-probability had features that determined the peculiarities of our own concept. The most important of these is entirely ignored in the conjectured

9

explanations discussed in this chapter. They forget that the probability emerging in the time of Pascal is essentially dual. It has to do both with stable frequencies and with degrees of belief. It is, as I shall put it, both aleatory and epistemological. This quite specific character of probability is one of the clues to its emergence. So now we must examine that duality with some care.*

---

* Another account of early ideas about chance appeared when this chapter was in proof: O. B. Sheynin, On the prehistory of the theory of probability. *Archive for the History of Exact Science* **12** (1974), 97–141.

2

DUALITY

According to legend probability began in 1654 when Pascal solved two problems and then wrote to Fermat. In fine detail this is wrong. The problems had been around for a long time and Pascal's chief clue to solution – the arithmetic triangle – is something Pascal might have learned at school and which was given in lectures even a century earlier. But like so many persisting legends the story of 1654 encapsulates the truth. The decade around 1660 is the birthtime of probability.

In 1657 Huygens wrote the first probability textbook to be published. At about that time Pascal made the first application of probabilistic reasoning to problems other than games of chance, and thereby invented decision theory. His famous wager about the existence of God was not printed until 1670 but it was summarized in 1662 at the end of the Port Royal *Logic*. The same book was the first to mention numerical measurements of something actually called 'probability'. Simultaneously but independently the adolescent German law student Leibniz thought of applying metrical probabilities to legal problems. He was also engaged in writing a first monograph on the theory of combinations. Also in the late 1660s annuities (long used by Dutch towns for financing public business) were being put on a sound actuarial footing by John Hudde and John de Witt. The London merchant John Graunt published in 1662 the first extensive set of statistical inferences drawn from mortality records. About the end of the same decade John Wilkins put forward a probabilistic version of the argument from design, prefacing his work with sentences like those made famous fifty years later by Joseph Butler: 'Probability is the very guide in life'.

In short, around 1660 a lot of people independently hit on the basic probability ideas. It took some time to draw these events together but they all happened concurrently. We can find a few unsuccessful anticipations in the sixteenth century, but only with hindsight can we recognize them at all. They are as nothing compared to the blossoming around 1660. The time, it appears, was ripe for probability. What made it ripe?

It is notable that the probability that emerged so suddenly is Janus-faced. On the one side it is statistical, concerning itself with stochastic laws of chance processes. On the other side it is epistemological, dedicated to assessing reasonable degrees of belief in propositions quite devoid of statistical background. This duality of probability will be confirmed by our detailed study of the history between 1650 and 1700. Even now it is clear enough. Pascal himself is representative. His famous correspondence with Fermat discusses the division problem, a question about dividing stakes in a game of chance that has been interrupted. The problem is entirely aleatory in nature. His decision-theoretic argument for belief in the existence of God is not. It is no matter of chance whether or not God exists, but it is still a question of reasonable belief and action to which the new probable reasoning can be applied.

The duality of probability is well illustrated by the list of the other contributors with which I have begun this chapter. Huygens wrote chiefly on aleatory problems. Leibniz began in an epistemological way, concerned with degrees of proof in law. When he came to Paris, he at once saw that the mathematics of 'Pascal, Huygens and others' fitted into his scheme. The Port Royal *Logic*, written by Antoine Arnauld and others, ends with a discussion of reasonable belief and credibility. Graunt's *Observations*, published in the same year, 1662, is entirely dedicated to demography and the analysis of stable frequencies. Yet the *Logic* has whole sentences of exactly the same form as are found in Graunt. Hudde and de Witt were doing the first actuarial science. Wilkins had no interest in that, and was concerned only with the probability of opinion. But they were both in the same new field, and were seen to be so by their contemporaries. Out of what historical necessity were these readily distinguishable families of ideas brought into being together and treated as identical? If there is such an historical necessity, it must be among the preconditions for probability. Hence we must not ask, 'How on the one hand did epistemic probability become possible and how, on the other, did calculations on random chances become possible?' We must ask how this dual concept of probability became possible.

Duality

The duality of probability has long been known to philosophers. The present generation may have learnt it from Carnap's weighty *Logical Foundations*. A century earlier one read about in, for example, S.-D. Poisson [1837, p. 31] or A. A. Cournot [1843, pp. v, 437–40]. Carnap said we ought to distinguish a 'probability₁' from a 'probability₂'; later he spoke of inductive and statistical probabilities. Poisson and Cournot said we should use the ready-made French words *chance* and *probabilité* to mark the same distinction. Before that Condorcet suggested *facilité* for the aleatory concept and *motif de croire* for the epistemic one [1785, p. vii]. Bertrand Russell uses 'credibility' for the latter [1948, p. 359]. There have been many other words. We have had *Zuverlässigkeit*, 'propensity', 'proclivity', as well as a host of adjectival modifiers of the word 'probability', all used to indicate different kinds of probability. The duality of probability is not news. Yet there is something wrong about it. Why these almost frantic gropings for a terminology to make distinctions? Consider an analogy. When Newton distinguished weight from inertial mass his terminology caught on at once. Just as credibility may sometimes be measured by frequencies, so under certain circumstances can mass be compared by weighing (at constant *g*). We do not thereby conclude that some construct based on the latin *pond-* is the word to be used for both. Why use *prob-* for both chance and credibility? We can understand why Baliani (in the preface to *De motu gravium* of 1638) might, while groping for Newton's distinction, speak of weight as *agens* opposed to weight as *patiens*. But once Newton's principles have been given in distinct terms, no one wanted to speak of active and passive weight. No one would have said that since both weight and mass satisfy the axioms of measure theory we need the same word for both. Yet that has been used as an argument for keeping the same word for aleatory and epistemic notions of probability.

It is true that the different kinds of probability are less well understood, and so less easily distinguished, than weight and inertial mass. By now, however, there are plenty of useful explications of the various probability ideas. On the epistemological side two schools of thought have been dominant. First, in the early decades of this century, there was much interest in the theory advanced by Harold Jeffreys and J. M. Keynes, according to which the probability conferred on a hypothesis by some evidence is a logical relation between two propositions. The probability of *h*, in the light of *e*, is something like the degree to which *h* is logically implied by *e*. Later students, despairing of a good analysis of this logical relation, have been more attracted to what L. J. Savage called 'personal probability', introduced by F. P. Ramsey and B. de Finetti. In this theory, the probability you assign to any particular proposition is a matter for your own personal judgement, but the set of all your probability assignments is subject to fairly strong rules of internal coherence. No matter whether the logical or personal theory be accepted, both are plainly epistemological, concerned with the credibility of propositions in the light of judgement or evidence.

In contrast there is a family of statistical theories, focussing on the tendency, displayed by some experimental or natural arrangements, to deliver stable long run frequencies on repeated trials. Some workers, following Richard von Mises, have attended chiefly to the phenomenal aspect of this, providing theories of randomness in infinite sequences and now, with results commencing with A. Kolmogorov and Per Martin-Löf, with randomness in finite sequences. Other students have come to think that the causes of frequency phenomena are more important than the phenomena themselves, and so, following Karl Popper, develop a concept of the propensity of a test of some sort to yield one of several possible outcomes. Clearly none of this work is epistemological in nature. The propensity to give heads is as much a property of the coin as its mass, and the stable long run frequency found on repeated trials is an objective fact of nature independent of anyone's knowledge of it, or evidence for it.

Philosophically minded students of probability nimbly skip among these different ideas, and take pains to say which probability concept they are employing at the moment. The vast majority of the practitioners of probability do no such thing. They go on talking of probability, doing their statistics and their decision theory oblivious to all this accumulated subtlety. Moreover there are a few extremists on either side. There are the personalists, including de Finetti, who have said that propensities and statistical frequencies are some sort of 'mysterious pseudo-property', that can be made sense of only through personal probability. There are frequentists who contend that frequency concepts are the only ones that are viable. Thus this labour of distinguishing kinds of probability has been curiously idle. Most people who in the course of their work use probability, pay no attention to the distinctions. Extremists of one school or another argue vigorously that the distinction is a sham, for there is only one kind of probability.

Carnap, and Cournot before him, notoriously failed to bring tranquillity out of controversy by their judicious mixture of conceptual analysis and linguistic distinction. Philosophers seem singularly unable to put asunder the aleatory and the epistemological side of probability. This suggests that we are in the grip of darker powers than are admitted into the positivist ontology. Something about the concept of probability precludes the separation which, Carnap thought, was essential to further progress. What?

There is an anti-positivist model which, for all its obscurity, may at this point have some appeal. We should perhaps imagine that concepts are less subject to our decisions than a positivist would think, and that they play out their lives in, as it were, a space of their own. If a concept is introduced by some striking mutation, as is the case with probability, there may be some specific preconditions for the event that determine the possible future courses of development for the concept. All those who subsequently employ the concept use it within this matrix of possibilities. Whatever the overall value of this strange model in the history of ideas, we can at least agree that since 1660 the concept of probability has been curiously autonomous and steadfast to its origins. In the past 300 years there have been plenty of theories about probability, but anyone who stands back from the history sees the same cycle of theories reasserting itself again and again.

Consider, for example, the most recent fashion, pioneered by F. P. Ramsey in 1926, and winning wide recognition after the book by L. J. Savage published in 1954. Some have called this subjectivism; Savage called it personalism. Most statisticians call it the 'Bayesian' theory. Thomas Bayes died in 1760, but the basic idea of his fundamental contribution is, two centuries later, the core of the most up-and-coming theory of statistical inference. Many a reader may begin to have a feeling of *déjà vu*.

Or again, consider the different interpretations that have been put upon the work of Jacques Bernoulli, whose posthumous book on probability was published in 1713. Because he introduced the word 'subjective' into philosophizing about probability, he has been called a subjectivist. Others say he anticipates Carnap's 'logicist' theory of probability. Still others call him the precursor of the extreme frequentist version of statistics of which Jerzy Neyman is the most famous living exponent. Neyman, Carnap, and subjectivism are all rightly recognized as virtually incompatible, yet different workers can, with some justice, trace their origins in the work of Bernoulli. Thus, at a very gross and as yet imperceptive level, we may readily confirm the fact that for all our advances in mathematical technology, a good many aspects of that dual concept of probability have been there from the beginning. The theories of today seem to compete in a space of possible theories that can be discerned even in the earliest years of our concept.

It is better to expose the crudities of one's model at the start, than to conceal a methodology in banal phrases. I am inviting the reader to imagine, first of all, that there is a space of possible theories about probability that has been rather constant from 1660 to the present. Secondly, this space resulted from a transformation upon some quite different conceptual structure. Thirdly, some characteristics of that prior structure, themselves quite forgotten, have impressed themselves on our present scheme of thought. Fourth: perhaps an understanding of our space and its preconditions can liberate us from the cycle of probability theories that has trapped us for so long. This last feature has a familiar ring. The picture is, formally, the same as the one used by the psychoanalysts and by the English philosophers of language. 'Events preserved in memory only below the level of consciousness', 'rules of language that lie deep below the surface', and 'a conceptual space determined by forgotten preconditions': all three have, of course, a common ancestor in Hegel.

I do not ask any reader to swallow all this. The story told in what follows is of interest even if the methodology that led to it turns out to be silly. I state the picture partly to explain how some of the data have been selected.

To begin with, the probability to be described is autonomous, with a life of its own. It exists in discourse and not in the minds of speakers. We are concerned not with the authors but with the sentences they have uttered and left for us to read. We do of course tag sentences with the names of authors, but that is largely a matter of convenience. This shall be particularly so in prehistory. We are not concerned with who wrote, but with what was said. This attitude will irritate the proper historian. He wants to know how an idea is communicated from one thinker to another, what new is added, what error deleted. I am more interested when the same idea crops up everywhere, on the pens of people who have never heard of each other.

My model has other implications. I tend to disregard the anticipation, the man who, with subtle interpretation, can be presented as a precursor of the modern way of thought. In prehistory we are not interested in what is rare but in what is common. Common does not mean familiar – it may be utterly bizarre. For example, I say, with only very slight reservations, that there was no probability until about 1660. How do I know? I have not read every text. There are many texts that no-one fluent in probability lore has ever read. How can I so confidently talk of the beginning of this family of ideas? Because I am talking about that time when this family emerges permanently in discourse. It is that event, and not the miraculous stroke of some secret hermit, that is our topic.

A third consequence of the model is that prehistory is quite severely localized. We are concerned with some substantial period of time before 1660. I shall glibly speak of 'the Renaissance' to mean, roughly, the fifteenth and sixteenth centuries, and try to steer clear of famous debates about what 'the Renaissance' was, if anything. We must focus on some aspect of the conceptual scheme of that time. It is true that we may find in Aristotle sentences translated as, 'the probable is what usually happens', but that was too long ago for us. Again, I shall much discuss the concept of sign, a good Stoic concept, much debated in, for example, Sextus Empiricus, and itself referring back to Aristotle. Those ancient sentences are not irrelevant, but Sextus matters to us not because of what he said, but only because of the use that Gassendi, say, made of him.

Enough of metatheory: it is chiefly an afterthought to characterize what one has been doing. Let us see what existed in the place of probability before 1660. The easiest place to start is with the word 'probability' itself, and to that I now proceed.

3

OPINION

Probability began about 1660, but the word 'probability' is, in languages drawing on the Latin, a good deal older. The prehistory of probability can usefully begin with a study of earlier meanings of the very word. Its link with numerical ideas of randomness seems first to have occurred in print only in 1662. Some English philosophers, stimulated by an interest in 'ordinary language', and perhaps suspicious of three centuries of success in making probability mathematical, have recently emphasized some pre-1662 aspects of the word. They go so far as to say that even today the primary sense of the word is evaluative. Thus according to W. C. Kneale, 'if we heard a man speak in ordinary life of the equal probability of various alternatives, we should understand him to mean that they are equally approvable as bases for action' [1949, p. 169]. Or again: in the 'common or garden' usage of the word 'probable', 'it is an evaluative term. To say that a proposition is probable is more like saying that it's right to do so and so' [Körner, 1957, p. 19]. Stephen Toulmin [1950], John Lucas [1970] and others have subsequently expressed similar views.

Undoubtedly the Latin word *probabilis* did mean, among other things, something like 'worthy of approbation', but I very much doubt if Kneale's account of the present 'common or garden' usage is right. One way to question it is to note how odd is the sound of older bits of speech in which 'probable' really did mean approvable. The impossibility of the old locutions shows how much the meaning has shifted, and it will also help to lead us back to even earlier senses of the word. A couple of centuries ago one readily spoke of a 'probable doctor', apparently meaning a medical man who could be trusted. We no longer speak that way. For a more striking example, consider a passage in Daniel Defoe's 1724 bawdy novel, *Roxana, or The Fortunate Mistress*. Early in her career the lady in question,

18

having got a man with a big house, says of herself, 'This was the first view I had of living comfortably indeed, and it was a very *probable* way, I must confess, seeing we had very good conveniences, six rooms on a floor, and three stories high.'

Since 'probable' had this connotation of approval it may seem reasonable to expect that when, in an antique work, the word is used to qualify some proposition, then the author is saying that the proposition is 'worthy of approval' because it has the marks of truth or is better supported by evidence than any other hypothesis. Such a conjecture requires much caution. Nowadays, according to J. R. Lucas, 'We use the words "probable", "probably", . . . to give a tentative judgement. There is some reason, but not conclusive reason, for what we opine.' Drawing attention to this claim, a reader of *The Times Literary Supplement* wrote to the editor [9 April 1971]:

it seems that for Gibbon in the eighteenth century [the words] had quite a different sense. Summing up a discussion of the conflicting accounts of Hannibal's route across the Alps in the ancient authors, he wrote in his journal for October 24 1763, 'Let us conclude, then, though with some remainder of scepticism, that although Livy's narrative has more of probability, yet that of Polybius has more of truth.' Still more surprisingly to a modern ear, he wrote in a footnote in Chapter xxiv of the *Decline and Fall*, 'Such a fact is probable but undoubtedly false.'

Such quotations may usefully shake up our preconceptions before we start a serious reading of the documents. Here is a final example. Throughout the first half of the eighteenth century there had been considerable controversy in Britain over the relation between miracles and testimony, and various pieces of probability lore were injected into the argument. In 1748 Hume made a particularly sensational attack on the credibility of miracles, based on his view of probability. This invited a host of serious replies, all of which convey some information about the current understanding of probability concepts. One of these books, by Thomas Church, considers the question of whether a fact can be credible or incredible in itself, 'distinct from the consideration of any testimony'. The author is at pains to insist that credibility is relative to the evidence. Church grants,

that in common discourse it is not unusual to call any thing credible or incredible, antecedent to our consideration of its proof. But if we examine our ideas, this will be found to be a loose unphilosophical way of expressing ourselves. All that can be meant is, that such a thing is possible or impossible, probable or improbable, or, at farthest, happening very frequently, or very seldom [1750, p. 60].

Here we see an array of all the concepts that have come to cluster around probability: credibility, frequency, possibility and the like. Probability is kept separate from each. Usages like those of Defoe, Gibbon and Church were at the end of the line, and began to die out as mathematical probability became more and more successful. They were the very standard earlier. Clearly we should not expect various formations of the word 'probability' in different languages naturally to translate into our own word. What then was the core of the preceding meanings?

E. F. Byrne has recently published a quite thorough study of probability concepts as they occur in the work of Thomas Aquinas, who is thus both a convenient and natural starting point. The first thing to grasp, as Byrne insists, is the distinction between knowledge and opinion in medieval thought. It is indeed an ancient distinction that we all associate with Plato, but here we are concerned with its manifestation in Aquinas. It contrasts strongly with all modern epistemology. For a good number of years, now, philosophers have been debating whether knowledge is justified true belief. Even when this brisk definition is rejected, it is very widely accepted that if *p* is true, then one person may believe *p* while another, in a happier epistemological state, may know the very same proposition *p*. Knowledge and belief are in the same line of business. If we are to understand the Thomistic doctrine we must adopt a different stance. Knowledge, so far from being justified true belief, does not even have the same objects as opinion.

In medieval epistemology, science – *scientia* – is knowledge. Knowledge is knowledge of universal truths which are true of necessity. The necessity in question is not identical to our concept called 'logical necessity' – a concept that did not properly exist until the seventeenth century. Aside from knowledge of first truths, which are so simple and fundamental that they are beyond disputation, knowledge is arrived at by demonstration. One of the requirements of Thomistic knowledge is that we have 'right' concepts. Take a simple example, the proposition that 'the plague is transmitted by fleas'. It is not possible to know this until at the very least we possess enough epidemiology to distinguish bubonic from pneumonic plague, and enough parasitology to distinguish the relevant kinds of flea. Some kinds of plague are transmitted by an organism that involves a kind of flea in its life cycle, and some are not. When we have that sort of understanding of pestilence, then we may begin to frame definitions that characterize the concepts in the scientifically relevant way. Of course it is contingent that those are the right concepts – the world might have been different. But once we do have an adequate theory of plague, it will determine the meanings of terms in such a way that it is at least plausible to say that by definition a particular kind of bubonic plague is transmitted by a particular member of the Siphonaptera – it is just that interrelation that is part of the characterization of the two species in question.

Let us now consider a resident of Cairo in 1837 who conjectured that the plague then infesting his city was transmitted by fleas. We may think more highly of him than his neighbour who attributed the plague to miasma. We have real respect for the Londoners of 1603 who blamed imported cotton for their plague: there is a particular parasite that needs a flea that needs cotton. But let us ask the question, whether the Egyptian of 1837 or the Londoner of 1603 believed something that we now know to be true. It is not unnatural to say that the propositions about the plague now accepted in epidemiology simply did not exist a century or more ago. Bubonic and pneumonic plagues had not, for example, been distinguished, and the very concept of a host parasite was undreamt of. There is a certain family resemblance between present knowledge and old opinions, but arguably no proposition central to modern theories of the plague is identical to any proposition believed a century ago. It is not the case that an old opinion, *p*, has become modern knowledge. The old opinion was in a sense incommensurable with modern knowledge. This usage of the term 'incommensurable' was popularized in the 1960s by Paul Feyerabend. Although the theory of scientific theories that accompanies it is rather at odds with recent positivism, it fits in quite well with some tenets of scholastic epistemology. There is, however, one fundamental difference. Aquinas thought that real infallible knowledge is a genuine goal and is sometimes attained. Knowledge is both distinct from and better than opinion. A Feyerabend might say instead that all our beliefs and theorizing are in the domain of opinion, and we should expect that any theory will have to be replaced by another one.

Aquinas' *opinio* refers to beliefs or doctrine not got by demonstration. It may also cover propositions which, not being universal, cannot (according to Aquinas) be demonstrated. *Opinio* tends to refer to belief which results from some reflection, argument, or disputation. Belief got from sensation is called *aestimatio*. In scholastic doctrine opinion is the bearer of probability. The limit of increasing probability of opinion might be certain belief, but it is not knowledge: not because it lacks some missing ingredient, but because in general the objects of opinion are not the kinds of propositions that can be objects of knowledge.

Even if we have apprehended the notion of opinion we are still far from medieval probability. We may expect that an opinion is probable if there are good reasons for it, or if it is well supported by evidence. This is not the primary sense that Aquinas attaches to probability, and it is instructive to see why. In his mind reason and cause are very closely related. To comprehend the reason for *p* is to understand the cause, to understand why *p*. Causes in turn are to be found in the real definitions that underlie the science. That is, all reasons are demonstrative, because causes are necessary causes. We have come to think that deduction is only one way of giving reasons, and that much evidence falls short of deduction. For the medieval, evidence short of deduction was not really evidence at all. It was no accident that probability was not primarily a matter of evidence or reason. Probability pertains to opinion, where there was no clear concept of evidence. Hence 'probability' had to mean something other than evidential support. It indicated approval or acceptability by intelligent people. Sensible people will approve something only if they have what we call good reason, but lacking an adequate concept of good reason Aquinas could handle only actual approval. Here is a typical statement about opinion and probability:

Since, then, the dialectical syllogism aims at producing opinion, the dialectician seeks only to proceed on the basis of the best opinions, namely what is held by the many or especially by the wise. Let us suppose, then, that one encounters in dialectical reasoning some proposition which could in fact be proven through a middle term but which on account of its probability seems to be self-evident. The dialectician needs no more than this [I. *Post. An.* 1. 38. n.258].

Aquinas continues by saying that 'in demonstration one is not satisfied with the probability of the proposition'. Probability requires probity and approbation but for demonstration we must be able to see and show what is what. The primary sense of the word *probabilitas* is not evidential support but support from respected people. Byrne has nicely summed up the elements of this concept:

Attribution of probability to opinion has various connotations. In the first place, it refers to the authority of those who accept the given opinion; and from this point of view 'probability' suggests *approbation* with regard to the proposition accepted and *probity* with regard to the authorities who accept it. In the second place, 'probability' refers to the arguments which are presented in favor of the opinion in question; and from this point of view it suggests *provability*, that is, capacity for being proven (though not necessarily demonstrated). In the third place, 'probability' takes on a somewhat perjorative connotation precisely insofar as the proposition in question is *merely* probable; for, from this point of view the proposition is only *probationary* and not strictly demonstrated as are propositions which are properly scientific [Byrne 1968, p. 188].

See how this sense of 'probability' survived into eighteenth century English. We have read Gibbon saying that something probable is false. In other words, an opinion commended by authorities is in fact wrong. He said Livy had more of probability but Polybius had more of truth. This meant that ancient and modern critics tend to weigh in on Livy's side, but in this case they are mistaken. When so understood Gibbon's usage is quite free from paradox. The usage 'probable doctor' and the like is restricted to just those sorts of professions where the layman must largely rely on the judgement of others. We can now get the savour of Defoe's *Fortunate Mistress*. Her agreeable town house 'is very probable indeed' – this means not that she approves it but rather that, in the esteem of her betters, this is a good leg up from her scruffy beginnings.

'Probability' chiefly meant the approvability of an opinion. This had a number of important consequences. One was the casuistical doctrine of probabilism which is the butt of the sixth of Pascal's *Provincial Letters* [10 April 1656]. Pascal is called the founder of modern probability theory. He earns this title not only for the familiar correspondence with Fermat on games of chance, but also for his conception of decision theory, and because he was an instrument in the demolition of probabilism, a doctrine which would have precluded rational probability theory. We must briefly discuss probabilism here, but the conception of probability as approval of opinion had a more important consequence. The Renaissance physicists were still dedicated to knowledge and demonstrative science. Hence we shall not find in their work any need for or serious use of probability concepts. The prehistory of

23

epistemological concepts lies in a less well known area, the purveyors of opinion. In particular, medical science had no hope of being demonstrative; nor even had the 'natural magic' which is the precursor of chemistry. It is in the probable signs of the physicians and the alchemists that we shall find the evolving concepts that make our kind of probability possible.

First a few superficial words on probabilism. It is a principle of casuistry advanced by the Jesuit order in the sixteenth century, and enjoying success, power, but great antagonism and finally defeat in the seventeenth. What is to be done when authorities, especially the Fathers of the Church, are found to disagree? The problem became pressing in the late Renaissance as more and more texts were discovered and more and more interpretations of existing texts were invented. Basically there are two possibilities. To resolve conflict, we can cut down on the authorities whom we will recognize, sticking only to scripture and the natural light of reason. Or we may consider a wide range of authorities, but in deciding among them, consider the social and moral effects of adopting their several doctrines. Roughly speaking the various protesting sects, including the Jansenists (who remained within the Church), took the former course, while the casuists took the latter.

Contrary to what is sometimes reported, probabilism in theology did not say that when authorities conflict, one should follow the most probable opinion. Probabilism says that one may follow some probable opinion or other, even a less probable opinion. The word 'probable' here does not mean well supported by evidence. It means supported by testimony and the writ of authority. When a doctrine is disputed, and you are in doubt as to how to act, you may, according to the probabilists, follow a course of action that is recommended by some authority, even when more or weightier authorities counsel the opposite course of action. But even that is only the half of probabilism. It tells what is permitted, but the Jesuits were not permissive. On the contrary, from the point of view of the Jansenists, the probabilists would first of all decide on a course of action for its social and moral expediency. Then they would find some old text that could be interpreted as approval of that course of action. Then, even if weighty authority tells one to do the very opposite, one may still proceed, for one is using a 'probable' opinion, namely an opinion that is authorized by someone or other.

Opinion

The Jansenist enclave at Port Royal included among its members
Antoine Arnauld (1612–94), Pierre Nicole (1625–95) and Blaise
Pascal (1623–62), who loom large in our history of probability.
Arnauld, perhaps the most brilliant theologian of his time, was
condemned by the Jesuits. The rivalry was old: in 1640 he had
printed a scathing little note on probabilism. After much politick-
ing, his enemies had him denounced, although the denunciation was
withdrawn in 1669. It had the effect of spurring Pascal to write
Provincial Letters by way of defence, or rather by way of attack. The
sixth letter is a reasoned repudiation of probabilism. A few rude
passages in the *Pensées* show how intensely Pascal detested that
casuist doctrine. Arnauld was reinstated by a council of 1669; at no
time had he lost the respect of those of his contemporaries whom we
most remember. He had participated in writing the 1662 Port Royal
Logic, which both contains an argument against probabilism and is
the first occasion on which 'probability' is actually used in what is
identifiably our modern sense, susceptible of numerical measurement. Throughout the rest of the century post-Jansenist writers
about probability occasionally took pains to say that they did not
have in mind the loathsome casuistical concept that bore the name
of probability. It is not to be inferred that the rise of probabilism had
nothing to do with the emergence of probability. Probabilism is a
token of the loss of certainty that characterizes the Renaissance,
and of the readiness, indeed eagerness, of various powers to find a
substitute for the older canons of knowledge. Indeed the word
'probabilism' has been used as a name for the doctrine that certainty
is impossible, so that probabilities must be relied on. I have written
above only of probabilism in theology, which is the doctrine that so
exercised our founders of probability. Such probabilism is still in
that medieval world where probability is an attribute of opinion,
and where probable opinion is that which is attested by authority. It
is not post-1660 probability at all, and, aside from political and
theological overtones, that is why the discoverers of the new
probability despised it so much.

I have said that we shall not find students of the physical sciences
making much use of anything they call probability, because they are
after knowledge, not opinion. Let us take for example Francis
Bacon (1561–1626) and Galileo Galilei (1564–1642). The former
was taken to be the philosopher of the new physics, and Galileo its
greatest practitioner. Now Galileo had, as we shall see, a good sense

25

of games of chance and was perhaps the first worker really to tackle the problem of how to make the best use of discrepant measurements of the same quantity. Had anyone seen that gaming and the theory of errors would merge with the old notion of 'probability', it should have been Galileo. But although the word *probabilità* occurs frequently enough in, say, the marvellous *Dialogue Concerning the Two Chief World Systems* [1632] it mostly has the old connotations. Indeed at one point Stillman Drake, the editor and translator into English, has to intervene with a footnote, ' "Not improbable" here means "not implausible, though incorrect".' Elsewhere Galileo called the opinion of Copernicus 'improbable' because of the plentiful experiences which overtly contradict the annual movement, and because of the strength in debate of the Ptolemaics and Aristotelians. 'There is no limit to my astonishment when I reflect that Aristarchus and Copernicus were able to make reason so conquer sense that, in defiance of the latter, the former became the mistress of their belief.' That is, Copernicus' opinion was improbable and still the one best supported by the deepest arguments. Here we may contrast Leibniz, writing less than a century later, taking this very same situation as one in which, despite all opinion to the contrary, the Copernican hypothesis was, at the time it was promulgated, 'incomparably the most probable'. For Leibniz probability is what is determined by evidence and reason; for Galileo, probability has to do with approval.

There do remain, however, excellent passages in which Galileo makes plain that approval ought to correspond to the evidence, not the weight of the authorities. For example, Sagredo asserts that the velocity of a body rolling down an inclined plane is a function of only the height of the plane. Salviati replies 'What you say seems very probable, but I wish to go further and by an experiment so to increase the probability of it that it shall amount almost to absolute demonstration.' The *esperienza* in question is based on a pendulum whose fall is arrested at various points. Ernst Mach maintains that it led Galileo to the law of inertia [1895, p. 143]. It does not seem to me that the argument in question does get anywhere close to 'absolute demonstration', but we here have a very clear indication of the notion that experiments – at least thought experiments – can increase probability almost to demonstration. There is no attempt to measure this increase in probability, nor is there any point in measuring it. Galileo longs for absolute demonstration. So did his chief contemporaries.

Opinion

As well as being the official philosopher of the new physics, Bacon is a good writer to turn to because as he says of himself in the *Novum Organum* [1620], he wants 'to banish all authorities and all sciences' – in particular dogmatic Aristotelianism and alchemical empiricism. Hence 'approval by the wise' is hardly going to be a means of appraisal, and the Latin or English word for probability will not refer to authoritative approval. But it still seems to mean 'worthy of approval', as for example in Sec. 122: 'With regard to the universal censure we have bestowed, it is quite clear to anyone who properly considers the matter, that it is both more probable and more modest than any partial one could have been.' It is no longer the wise who confer probability by their approval; it is those who properly consider the matter. If one does not consider the matter properly, things may only 'seem probable':

The empiric school produces dogmas of a more deformed and monstrous nature than the sophistic or theoretic school; not being founded in the light of common notions (which, however poor and superstitious, are yet in a manner universal, and of a general tendency), but in the confined obscurity of a few experiments. Hence this species of philosophy appears probable, and almost certain to those who are daily practiced in such experiments, and have thus corrupted their imagination, but incredible and futile to others [Sec. 64].

In short, if one does nothing but wretched experiments, opinions will appear probable which can hardly be approved by someone who has a broader stance. Note Bacon's dedication to the scholastic conception of knowledge. Knowledge is derived from common notions and states only universal truths.

Our course and method however (as we have said, and again repeat) are such as not to deduce effects from effects, nor experiments from experiments, (as the empirics do) but in our capacity as legitimate interpreters of nature, to deduce causes and axioms from effects and experiments; and new effects and experiments from those causes and axioms [Sec. 117].

The Baconian doctrine is not unlike what has come to be called the hypothetico-deductive method in science, except that there is that residue of the Middle Ages that later generations found pernicious: we seek true axioms and real notions that will ultimately produce knowledge and not opinion. There is little room in this conceptual scheme for a working concept of probability. Readers of Bacon or Galileo in the latter half of the seventeenth century found them the great originators of the new experimental method combined with a successful mechanical model of the universe. It has only quite recently been recognized that this interpretation is an artefact of the period after 1650, particularly among members of the Royal Society of London. If we examine the texts of Bacon or Galileo we find a world of first causes. There is no need here for a mathematical concept of probability, nor even a real use for qualitative probabilities. It is not to the 'high sciences' of astronomy, geometry, and mechanics that we must look. Instead it is those lowly empirics who had to dabble with opinion.

Opinion is the companion of probability within the medieval epistemology. There is another concept of equal importance to those empirics who had to work with opinion. This is the *sign*. Inevitably Shakespeare records it : 'The least of all these signs were probable' [Henry VI. 2. 78]. Leibniz, in running over the prehistory of probability, is chiefly attentive to the law – see Chapter 7 below – but recalls how 'the physicians have the various signs and indications which are in use among them'. [*P.S.* v, p. 447] The history of the concept of a sign is of fundamental importance. In the medical textbooks of the Renaissance there is a characteristic distinction between cause and sign. The causes are chiefly efficient causes, namely what make the person ill, and the signs are not so much what we might call symptoms, as anything by which we may make a prognosis. To take an example almost at random, H. von Braunsweig in 1574 is saying that 'When a man hath a great disease or feebleness and a cold sweat breaketh out only about the nose, that is a very deadly sign.' That sounds familiar enough, but we will also find something else. Here I quote from Fracastoro (1483–1553) on contagion, to whom is often attributed the first germ theory of disease:

Contagions have their own peculiar signs of which some announce beforehand contagions to come, while others indicate that they are already present. The signs that are called premonitory come from the sky or air or from the vicinity of the soil or water, and among these some are almost always, others are often, to be trusted. Therefore one ought not to consider them all as prognostications, but only as *signs of probability* [1546, Bk. I, Ch. xiii].

The signs in question are a heterogenous collection: the planets in conjunction, frequent comets, tempest flares from unctuous foams, and mildew on drying linen when the wind blows from the East.

Swarms of locusts intrigue the author, and once he bursts into verse: 'Often a tiny mouse shall give thee augury of ill. No tie of love can hold it beneath the depths of the earth but it breaks forth from its trenches, forgets its life and its habits and leaves its tender young and pleasant abode . . . .' The swarms of mice that occasionally overran some of the towns of Central Europe, thousands dying frothing in the streets, were indeed a probable sign of plague to come. However, it is of no moment which signs seem sensible to us, and which absurd. Here we have a very clearly stated conception of partial prognostication, which is thereby possessed of probability, rather than certainty, and whose probability arises from frequency, from what happens 'almost always' or else 'often'.

It is important that Fracastoro is not a mere empiric of the sort castigated by Bacon. One of the fundamental features of the new science of the seventeenth century was the distinction into primary and secondary qualities. Philosophers know this through Locke, and so miss the point of the distinction. The problem was to make a science – in the Scholastic sense of the term – out of alchemy. The solution, made permanent by Robert Boyle, was to insist that the phenomena of chemistry were to be explained by noumenal things in themselves, little bouncing particles, moving, but not coloured, collectively taking up space but not in themselves having taste. It was for a long time an excellent model. It was hardly new with Bacon or Boyle. As Fracastoro put it, the qualities 'that are called primary generate and alter everything, but those that are called secondary, namely light, smell, taste and sound, do not act on one another but merely serve to arouse the senses' [Ch. vi]. Thus in the domain of causation we will have a set of universal propositions involving primary qualities only. Knowledge of this is knowledge of how the world works; it is science. However, at the level of phenomena there is something else. When the patient comes to Fracastoro, he is blotched, stinks, complains of a foul taste in his mouth and sounds strange when thumped on the back; above all he complains of pain. The causes of all this lie inside the patient and are ultimately to do with atoms. But the signs are all secondary qualities, and in these signs we have to make merely probable prognoses. The real world is described by universal truths, but the Renaissance physician has to prescribe and predict from the phenomena. Our Galileo or Bacon could pursue the real world constantly seeking demonstration, but our Fracastoro must make

29

prognoses on the basis of what phenomena follow what with greatest frequency.

The connection between sign and probability is Aristotelian. 'Sign', however, had a life of its own in the Renaissance, to our eyes a bizarre and alien life, but a life that we must understand if we are to comprehend the emergence of probability. The old probability, as we have seen, is an attribute of opinion. Opinions are probable when they are approved by authority, when they are testified to, supported by ancient books. But in Fracastoro and other Renaissance authors we read of signs that have probability. These signs are the signs of nature, not of the written word. Yet we shall see, in the next chapter, that this antithesis is wrong. Nature is the written word, the writ of the Author of Nature. Signs have probability because they come from this ultimate authority. It is from this concept of sign that is created the raw material for the mutation that I call the emergence of probability.

4

EVIDENCE

Many modern philosophers claim that probability is a relation between an hypothesis and the evidence for it. This claim, true or false, conceals an explanation as to the late emergence of probability: the relevant concept of evidence did not exist beforehand. The way in which it came into being has much to do with the specific way that the dual concept of probability emerged. One of the preconditions for probability was the formation of this concept of evidence.

What concept of evidence? Crudely, that which some philosophers have called 'inductive evidence'. The label is inaccurate, but at the beginning it can remind us of the philosophers' problem of induction, almost always attributed to David Hume's *Treatise*, published in 1739. Some elements of this problem may have been anticipated in the *Outlines of Pyrrhonism* [ii, 204], written by the Greek sceptic, Sextus Empiricus (*c.* A.D. 200). But aside from odd and fragmentary passages almost certainly dedicated to other problems we find no hint of a problem of induction until Hobbes, or, better, Joseph Glanvill's *Vanity of Dogmatizing* of 1661. All modern students of epistemology agree that the problem of induction is of fundamental importance. Most of the other basic problems can be identified throughout the whole Western tradition, and have classic texts in Plato or Aristotle. Why is what C. D. Broad called the scandal of philosophy – the problem of induction – such a new-comer on the scene? There is a simplistic answer. Until the seventeenth century there was no concept of evidence with which to pose the problem of induction!

There are defects in this answer. First, despite such intimations as one may find in Glanvill in 1661, it is significant, and explicable, that the problem of induction had to wait in the wings some eighty years after the birth-decade of probability. As I shall explain in Chapter 19, Glanvill merely raises the flag over a new philosophical continent, discovered at the time of probability, but which cannot be exploited until other events have occurred. But our simplistic answer is partly right. A concept of evidence is a necessary condition for the stating of a problem of induction. A problem of induction does not occur in the earlier annals of philosophy because there was no concept of evidence available.

'Evidence', however, is far too imprecise a term. Of course some concepts of evidence have been around for a very long time. In this chapter I propose to define one concept of evidence which, I claim, was lacking. In the next chapter I shall describe the terms in which it came into being. My definition of this concept of evidence must, of necessity, be by way of exclusion. I shall describe a number of different kinds of evidence that were not lacking, and label these in various ways. What all of these leave out is something like what our philosophers have come to call 'inductive evidence'.

Concepts of testimony and authority were not lacking: they were all too omnipresent as the basis for the old medieval kind of probability that was an attribute of opinion. Testimony is support by witnesses, and authority is conferred by ancient learning. *People* provide the evidence of testimony and of authority. What was lacking, was the evidence provided by *things.* The evidence of things is not to be confused with the data of sense, which, in much modern epistemology, has been regarded as the foundation of all evidence. On the contrary, we should be concerned with that kind of evidence that J. L. Austin has nicely distinguished from sheer looking:

The situation in which I would properly be said to have *evidence* for the statement that some animal is a pig is that, for example, in which the beast itself is not actually on view, but I can see plenty of pig-like marks on the ground outside its retreat. If I find a few buckets of pig food, that's a bit more evidence, and the noises and smell may provide better evidence still. But if the animal then emerges and stands there plainly in view, there is no longer any question of collecting evidence; its coming into view doesn't provide me with more *evidence* that it's a pig, I can now just *see* that it is [1962, p. 115].

The evidence that will concern us, then, is not the 'evidence of the senses'. In Austin's examples, it is the evidence of things, such as the pig bucket, and perhaps also the noticeable noises and smells. These olfactory and auditory objects are not private experiences but rackets and stenches as public as pigsties.

The evidence of things is distinct from testimony, the evidence of witnesses and of authorities. Probably Austin did not mention witnesses because they seem parasitic on the evidence of things. We rely on them when we can not be at the scene ourselves. We use authorities when we are ignorant. People and books, whether they be authorities or chance witnesses, seem to stand in place of ourselves. They report on evidence that they have been able to acquire, and so it seems to us that they are not the basic kind of evidence. The Renaissance had it the other way about. Testimony and authority were primary, and things could count as evidence only insofar as they resembled the witness of observers and the authority of books.

Our form of the distinction between these two kinds of evidence, testimony and the evidence of things, is quite recent. It was clearly stated in 1662, at the end of the Port Royal *Logic*. The authors call the evidence of testimony *external* or extrinsic. The evidence of things is called *internal*. One may find this distinction a few years earlier, for example in Hobbes, but it is, in the hands of these authors, a new distinction. It is our distinction, and characterized in a way that we understand: the primary evidence, the evidence of things, is 'internal', and thereby basic, while testimony is 'external'.

I claim not only that the distinction is new, but also that the very concept of internal evidence was new. Internal evidence must not be confused with verisimilitude. We say that a proposition has verisimilitude when it is a proposition of the sort that is true. For example, when in 1440 Lorenzo Valla (*c.* 1406–57) exposed the fraudulent Donation of Constantine, he did so in a way that modern textual critics find very strange. Indeed, as one of these has remarked to me, 'he did not use any *evidence* at all!' Lorenzo instead considered whether the Donation is the sort of thing that could have happened. Constantine, according to documents, donated the Roman Empire to the Church after his miraculous cure from leprosy. Lorenzo imagines a long conversation between Constantine, giving the Empire to Pope Sylvester, and Sylvester declining. No Emperor would ever give away his dominion, nor any Vicar of Christ accept it. And look at the very prose, continues Lorenzo: it is not the sort of thing to occur in an historical document.

Modern textual critics take solecisms and historical anachronisms as evidence that a text is faulty or fraudulent. That is a case of one thing (these particular words) serving as evidence against the claim that the whole text is sound. Just like Austin's pig-food, they are instances of one thing being evidence for another. We can recognize the production of some evidence in Lorenzo's polemic, but Lorenzo himself is not arguing that way. He is saying that this document is not like a true document: it lacks verisimilitude. Evidence, in my usage, is a matter of inferring one thing from another thing, while verisimilitude is a matter of one thing being, or not being, what it seems or pretends to be.

The kind of evidence that I have in mind consists in one thing pointing beyond itself. This must be further clarified. It is non-deductive pointing. A single observation that is inconsistent with some generalization points to the falsehood of the generalization, and thereby 'points beyond itself'. But this pointing is by way of *reductio ad absurdum*, a demonstrative form of argument. Such form of argument was well known to the *scientia* of medieval times and the early Renaissance. Here is a typical example, by the Archbishop of Canterbury, John Pecham (*c.* 1230–92).

Proposition 28: *Sight occurs through lines of radiation perpendicularly incident on the eye*. This is obvious, for unless the species of the visible object were to make a distinct impression on the eye, the eye could not apprehend the parts of the object distinctly. [Lindberg 1970, p. 109].

This is from a manuscript which, under the name *Perspectiva Communis*, circulated widely in the fourteenth century. Whether or not the argument be persuasive, the form of the argument seems plain enough. Sight occurs through lines of sight perpendicular to the eye, or it does not. We have a known fact inconsistent with the latter, so the former must be true.

Demonstration, testimony and verisimilitude were quite well understood at the beginning of the Renaissance. Only internal evidence was lacking. Now to say that there was no concept of internal evidence is not to say that people did not use what we call evidence. Doubtless men have long inferred that there was a pig in the thicket from the sound, smell, and broken branches. But dogs and boars can tell there is a pig, and do not thereby have a concept of evidence. We do not deny that men in the Renaissance were able to take advantage of what we call the evidence. I deny that their description of this practice was at all like our description, or even fits into any present category.

Naturally I here make no claim about Sanskrit or Greek concepts of evidence. I am concerned with a specific lack at a particular time, and am interested in what stood in place of evidence. This, as we shall see, was the 'sign'. What happened to signs, in becoming evidence, is largely responsible for our concept of probability. We cannot even speculate about how another concept of probability might have emerged elsewhere at another time, from the transformations in another culture.

It will be my claim, in the next chapter, that the concept of internal evidence of things is primarily a legacy of what I shall call the low sciences, alchemy, geology, astrology, and in particular medicine. By default these could deal only in *opinio*. They could achieve no demonstrations and so had to resort to some other mode of proof. The high sciences, such as optics, astronomy, and mechanics, still lusted after demonstration and could, in many cases, seem to achieve it. They could scorn *opinio* and any new mode of argument. New modes of argument arose, perforce, among the students of opinion. I shall be using some of the more bizarre examples taken from the hermetics because they so forcefully illustrate what seems to me to be important, but we can find exactly the same emergence of the 'sign' and the new kind of evidence in the sane and cautious words of the geologist Agricola (1490–1555) who remained in the established cloisters, as we shall find in the drunken speculations of the itinerant physician Paracelsus (1493–1541).

Before proceeding to the study of signs, I should make a distinction between evidence and experiment. There is an ongoing debate among historians of science as to the roots of the 'experimental method'. Some historians attribute the method to the growing self awareness of the new mechanics. Their chief hero is Galileo, a man who, even if he did not experiment as much as was once thought, admired and imagined many experiments. Other historians emphasize the role of the low sciences, emphasizing the bizarre laboratories of the new physicians and alchemists. Yet a third school of history claims that there are different experimental traditions that converge in the seventeenth century. Since I shall be discussing the origin of the concept of something like 'inductive evidence' it may seem as if I can contribute to this debate about origins, but that impression is largely illusory.

To begin with, we may distinguish, abstractly, numerous kinds of experiment. I shall call them, for ease of reference, the test, the adventure, the diagnosis, and the dissection. The *dissection* is a matter of taking something apart to see what is inside. It has a primarily visual motivation. The early dissections of Vesalius and his peers have been much studied in the history of science, although undoubtedly the more recent positivist thesis, that seeing is believing, has distorted our understanding of what was once done in the dissecting room. The *test* is entirely different, and operates by that inner seeing which is deduction. One tests an hypothesis *H* when *H* implies that if event *E* occurs, then result *R* will follow. One endeavours to make *E* occur. If *R* fails to follow, then *H* is confuted. If *R* does follow, *H* is thereby corroborated. We have come to think of a positive result of a test as somehow conveying inductive evidence for *H*, but that was not the original intention, for there was no concept of inductive evidence. Passing the test was often called a proof of *H*. Here proof bears that old sense we still find in expressions like 'printers' proofs' or, 'the proof of the pudding is in the eating'.

The test is conducted in circumstances where, if one believes the theory, one has firm expectations about the outcome. An *adventure*, in contrast, is guided by no good theory and we may only guess what will happen. Much early alchemy seems to have been adventure. You heated and mixed and burnt and pounded to see what would happen. An adventure might suggest an hypothesis that can subsequently be tested, but adventure is prior to theory.

An adventure is an end in itself. Indeed, the ultimate aim may be to make gold or to find out more about the universe, but the adventure is done for its own sake. To this we contrast the *diagnosis*. In a diagnosis, for example, you add substances to the urine of a sick man, collect the precipitate and pound it. Perhaps you can only guess the outcome, but this is not a pure adventure. Rather, from the character of the precipitate you infer what is wrong with the patient. The surgeon cuts up live people and the anatomist dissects the dead, but the physician must be content with reading the signs in his laboratory.

Tests, adventures, dissections and diagnoses all provide 'evidence'. The evidence that they provide is of differing kinds. The test demonstratively refutes an hypothesis, or else corroborates it. The adventure suggests a theory. The dissection exhibits the inner working of man and beast. My preceding discussion has excluded all these kinds of evidence. The Middle Ages possessed a concept of each kind of evidence provided by such experiments. Only the diagnosis gains, in the Renaissance, a new conceptualization. It uses a thing, the precipitate, as evidence for another thing, the state of man's insides. It is not a matter of simply looking, nor a matter of testing, nor a matter of guessing a new law in the light of an adventure. It is the evidence of one thing that points beyond itself.

The 'experimental method' is truly of many kinds and has many origins. The internal evidence of things need not be conceptualized before there is experimental method. The diagnosis has not that much to do with the origin of the experimental method. It may, however, have something to do with the interpretation in the seventeenth century, when 'the experimental method of reasoning' became exalted above all else. It became fashionable to regard all experiment as what I have been calling diagnosis. In the old Aristotelian tradition *scientia* was to proceed by the demonstration of effects from first causes. In the new science, one was to infer the causes from experiment. The old causes got at the essence of things. The new causes were efficient causes, explaining how things were made to work. You inferred the efficient causes from experiment. You inferred something small, inner, atomic, and precise from something, large, outer, gross and inaccurate. Just as the physician read the state of his patient from the signs in the urine, so the scientist was supposed to read the state of the atomic world from his crude diagnostic tools. In this way the test, for example, was transformed. The tests of the old *scientia* were demonstrative, and the result of passing a test was just that: passing a test. But in the new philosophy of the inductive sciences the result of passing a test was to get new inductive evidence for the hypothesis. One was, as it were, diagnosing the good health of the hypothesis. Karl Popper's methodology of science, brilliantly expounded in his *Logic of Scientific Discovery,* is an attempt to cast out from science the alchemists, the physicians, and their diagnostic experiments, re turning science to a plain demonstrative model.

We can here better understand a certain ambiguity in the philosophers' term of art, 'inductive evidence'. It has come to mean two things. On the one hand evidence for a generalization or even for a law of nature, gained from particular observation and experiment. On the other is the induction from particular to particular. Hume, in fact, chiefly considers the latter, as when he wonders whether *this* piece of bread before me is nourishing. J. S. Mill went so far as to claim that all inference is from particulars to particulars, generalizations being merely the schema of particular inference. In the Renaissance the evidence of particular things for particular things emerged first. The 'proof' of generalizations earlier used deductive modes of inference, as in my quotation from Pecham. When all experiments began to be conceived of as diagnosis, one was no longer diagnosing the state of the hidden liver, but rather the hidden laws of the universe, and so inductive inference for generalizations, and induction from particulars to particulars, become conceived of as in the same line of business.

Thus I do admit that my thesis on the origin of the concept of evidence may connect with current debates on the experimental method. This is not because our low scientists were peculiarly experimental, but because one kind of experiment in which they engaged had much to do with the subsequent interpretation of all post-Aristotelian science. Doubtless the technology devised by the proto-chemists affected what men did, but the true effect, of lasting importance to the new civilization, may lie in how men thought about what they did. Probability and the new understanding of experiment both had as their preconditions a transformation of an old concept of sign into a new concept of evidence. That we must now describe.

5

SIGNS

To understand the new kind of evidence delineated in the preceding chapter we must not look at the physicists competing for demonstrative knowledge but at the purveyors of opinion whom I have called the low scientists. The early empirics whom Francis Bacon so denigrated were chiefly alchemists, astrologers, miners and physicians. Every man endowed with lively curiosity pursued every trade, so there is no sharp division into high and low. Cardano, the author of the first book on probability, was famed both for his skill in medicine and his talent at mathematics, but for all the breadth of his interests he can safely be called a student of the low sciences. Copernicus, well versed in medical lore, was a high scientist. However we may quarrel about individuals we can often allot a given piece of work to one category or the other.

Herbert Butterfield has rightly warned that scholars who try to theorize about alchemy 'become tinctured with the kind of lunacy they set out to describe [1957, p. 129]. If we could study the high science of the Renaissance – Copernicus, say – we might stay quite sane. But probability emerges from low science. In recounting the work of the empirics it is of no value sedately to say that they combine science and occultism, and then leave out the 'occult'. We must instead try to absorb an alien conceptual scheme. We must try to comprehend a science,

in which there are two kinds of operation, one produced by nature itself, in which there is a selected man through which nature works and transmits her influence for good or evil, and one in which she works through other things, as in pictures, stones, herbs, words, or when she makes comets, similitudes, halos and other unnatural products of the heavens.

These are the words of Paracelsus [*Werke*, XII, p. 460]. In his own time (1493–1541) he was called 'the Luther of the Physicians'. In the next era John Donne was to describe him in verse as a greater revolutionary than Copernicus. Yet in the mind of Paracelsus that strange array at the end of my quotation – *bilder, stein, kreuter, wörter* [. . .] *cometen, similitudines, halones und ander des gestirns unnatürliche generationes* were all what modern philosophy calls a 'natural kind', namely a collection between which there are manifest family resemblances. The resemblances between words and stones, herbs and comets, are now lost to us, yet it is the conceptual scheme engendering such resemblances that we must try to penetrate. These are not the idle groupings of a man not given to distinctions: 'The physician must know that there are a hundred, indeed more than a thousand, kinds of stomach', says Paracelsus with contempt of those who have a single panacea for all stomach ache [VI, p. 153]. Nor is he an uncritical reciter of tales:

I do not compile my textbooks from excerpts of Hippocrates or Galen. In ceaseless toil I create them anew, founding them upon experience. If I want to prove anything I do so not by quoting authorities but by experiment and reasoning [Sudhoff 1894, I, p. 29].

Paracelsus is a convenient focus for our study of signs. His biographers portray him as a strange figure, a trifle more bizarre than many another of the hermetical wandering physicians who could serve as a model for Faustus. Modern histories of medicine acknowledge him as the man who brought chemistry into medicine, treating patients not only with herbs and seeds but also with distillates and precipitates. It is remembered that he challenged Galen's theories of antipathy, treating diseases by similar substances rather than by opposed ones. His new theory of the elements – mercury, salt, and sulphur – was a great spur to chemistry. But otherwise, in standard histories, his place is incomparably smaller than his fame in, say, 1600. He is a figure of the age who was revered as a great man by several succeeding generations and then almost forgotten.

The high sciences of the Renaissance have received much scholarship, but only now is low science being studied. There is some debate as to its role in the formation of European thought. Here our concern is not with the general issue but only with the notion of sign. Its structure begins with a truism which my last quotation from Paracelsus will have recalled: in the early Renaissance, books were too much revered. There is undoubtedly more to the veneration of ancient manuscripts than mere respect for a newly discovered classical culture, but that is not our topic. Rather we are concerned with the transformation from the study of books to the study of nature. Notice in passing how perfectly the constant copying and commenting ties in with the probability of *opinio*. 'Probable' meant 'approved by the wise'. If we follow the exhortation to write down only what is probable or demonstrated, then, in that old sense of 'probable', it is an analytic truth that we should recopy the words of others. He who strives after probable opinion can, of necessity, be only a copyist and a commentator.

Paracelsus and a thousand other voices came to protest the vain repetition of Galen, Avicenna and the like. But they did not say, let us abandon this external evidence and proceed to internal evidence. They did not say, give up copying and look at the facts. Rather they said, stop studying bad books and start studying good ones. 'How can the unlearned man be led out of ignorance to science?' – 'Not through your books, but God's.' – 'Which are they?' – 'Those which he wrote with his own fingers.' – 'Where are they to be found?' – 'Everywhere.' Nicholas of Cusa (1401–64) wrote that in his dialogue *Idiota* [1967, p. 217]. Long before the birth of Paracelsus the radicals were rejecting the commentaries. The greatest rejection was that of Martin Luther. But Luther did not invite us to give up book-learning. He inveighed against vain testimony, and told us to get back to The Book, to the real testimony,

The Renaissance did indeed struggle to liberate itself from the written word and take up the study of nature by experiment. But the revolutionaries saw themselves as returning to the words that really have been written. Here is Paracelsus:

The first and highest book of medicine is called *Sapienta*. Without this book no one will achieve anything fruitful [. . .] for this book is God himself [. . .] The second book of medicine is the firmament [. . .] for it is possible to write down all medicine in the letters of one book [. . .] and the firmament is such a book containing all virtues and all propositions [. . .] the stars in heaven must be taken together in order that we may read the sentence in the firmament. It is like a letter that has been sent to us from a hundred miles off, and in which the writer's mind speaks to us [XI, 171–6].

Many readers will suppose that Paracelsus speaks metaphorically of books, sentences, letters, alphabet and reading. He does indeed speak of 'reading the urine', it will be protested, but so does last week's brochure giving instructions on how to use pregnancy-testing equipment. The brochure speaks metaphorically. Why not Paracelsus too? The answer is twofold. First, because he himself makes no distinction. Second, and more important, because the literal sense of his words is essential to the sense of his system. To see that we must go a little deeper into his scheme of things.

It is well known how Galen ran medicine on the principle of the mean. Afflictions must be treated by contraries. Hot diseases deserve cold medicine and moist illnesses want drying agents. Treat excess of y by something deficient in y and thereby restore the balance. Paracelsus rebelled; he said that we must treat by similarity and not by difference. To cure a large dose of poison treat with a tiny dose of the same poison. To cure the liver treat with a herb that is shaped like a liver. He liked to quote Hippocrates' claim that experiment is futile. Quite so he said, in Hippocrates' time, 'but now we have a theory!' Since we know what sorts of medicines will be good for what sorts of ailments, we are able to begin to experiment with precisely measured doses.

Any theory that treats disease by similarity will require a theory of similarity. Paracelsus has that. It is the *doctrine of signatures*. Each thing has a signature and the physician must master the signatures. Signatures are ultimately derived from the sentences in the stars, but a bountiful God has made them legible on earth. Everything is written. Nature

indicates the age of a stag by the ends of his antlers and it indicates the influence of the stars by their names. Thus she made liverwort and kidneywort with leaves in the shape of the parts she can cure [. . .] Do not the leaves of the thistle prickle like needles? Thanks to this sign the art of magic discovered that there is no better herb against internal prickling [XIII, pp. 376–7].

In our conceptual scheme the names of the stars are arbitrary and the points on the antler are not. For Paracelsus both are signs and there are true, real, names of things. He often rants against his contemporaries and the ancients who called things by their wrong names, having forgotten, perhaps in Babel or at The Fall, what the names really are. For example, Paracelsus knew that the metal mercury, in the correct dosage, would cure syphilis, and he thereby established medical practice for three centuries. He knew this despite the fact that his colleagues were killing their patients by randomly treating syphilitics, among others, with mercury. Syphilis is signed by the market place where it is caught; the planet Mercury has signed the market place; the metal mercury, which bears the same name, is therefore the cure for syphilis.

The sign was a matter of reading the True Book. Bizarre hermetics like Paracelsus tell us so, but we do not need to consult them exclusively. It is a relief to get back to sober instruction, for example as furnished by Georgius Agricola in *De re metallica* [1556]. His method of reading the signs on the surface of the earth will (we feel) surely help the miner and the entrepreneur for whom the book is written. We cheer his cautious criticism of the alleged merits of divining ore from twitches of hazel sticks. This is a man who understood evidence. He is one of us, it seems, and Paracelsus seems quite alien. Yet when we look again we find that Agricola too is telling us how to read aright, and how to find the Sentences on the earth's surface that say what minerals are down below. We must accept that Agricola (born 1490) and Paracelsus (born 1493) use the same concepts although they have different styles. Nor is this a phenomenon of the 'German renaissance'. In Padua, the intellectual capital of the world, we found Fracastoro (born 1483) telling us that 'the earth itself shall give thee signs', 'as though she knew what is to come, as she quakes and sighs issue from her entrails'. When the world gave a sign of *p*, it attested to *p*. Hence in the old sense of 'probable', *p* was probable. The proposition *p* was not probable in our sense of the word, as having much evidence of experience in its favour. It was probable in the old sense of the word, as being *testified* to by sound authority.

When, however, are signs to be trusted? For although a reading of the book of the universe, if complete, would always be trustworthy, we have not yet managed to read the one great sentence that is writ upon the firmament, and must rely on the microcosm around us. Not all signs are equally trustworthy. As Fracastoro put it, 'Some signs are almost always, others are often to be trusted', and these are 'signs with probability'. It is here that we find the old notion of probability as testimony conjoined with that of frequency. It is here that stable and law-like regularities become both observable and worthy of observation. They are a part of the technique of reading the true world.

In Chapter 2 I emphasized the duality of the probability that emerged around 1660. On the one hand it is epistemological, having to do with support by evidence. On the other hand it is statistical, having to do with stable frequencies. Any theory on the emergence of probability must try to explain why the concept that emerged was dual in just this way. The old medieval probability was a matter of opinion. An opinion was probable if it was approved by ancient authority, or at least was well testified to. This medieval concept of probability is indeed related to our own, but in a surprising way. A new kind of testimony was accepted: the testimony of nature which, like any authority, was to be read. Nature now could confer evidence, not, it seemed, in some new way but in the old way of reading and authority. A proposition was now probable, as we should say, if there was evidence for it, but in those days it was probable because it was testified to by the best authority. Thus: to call something probable was still to invite the recitation of authority. But: since the authority was founded on natural signs, it was usually of a sort that was only 'often to be trusted'. Probability was communicated by what we should now call law-like regularities and frequencies. Thus the connection of probability, namely testimony, with stable law-like frequencies is a result of the way in which the new concept of internal evidence came into being.

The concept of sign as evidence, with its attendant implications of testimony, reading, and probability became the standard in all walks of life. Perhaps it is possible to see this as part of the intellectual back-sliding and obscurantism that is sometimes attributed to Renaissance thought. Some historians tell us that the high middle ages were full of 'good science' that gradually ran downhill in the fifteenth and sixteenth centuries. I think that the truth in their assertions is that *scientia* ultimately modelled on Aristotelian canons is collapsing and *opinio* is still formulating its own kind of evidence. Be that as it may, I do not claim that the concept of sign-as-evidence is a 'progressive' notion. We note only that it occurs in more and more of the sentences, written down in those days, and which have been preserved.

The sign-as-evidence indicates with probability but I do not claim that the authors who employed it were an 'influence' on the founding fathers of probability. Some historians of ideas are much concerned with the way in which work of *A* can influence his successor *B*. Two kinds of influence are considered. *B* may deliberately choose to employ central concepts or techniques of *A*, or else *B* may unwittingly pursue a programme initiated by *A*. Such talk of 'influence' is part of the historians' language of precursors and anticipations. It would be amazing if a Paracelsus were an 'influence' on a Pascal or a Leibniz. The mathematicians despised what they knew of the occult. Yet their contempt for those earlier hermetical figures does not preclude the possibility that whenever these geometers thought about opinion, they thought in a conceptual space that was the legacy of the very empirics whom they scorned. The intellectual objects about which, and *in* which, the new mathematicians thought had been formed in the crucibles of the alchemists and the vials of the physicians.

To prove this we must ourselves stop speculating about preconditions and briefly examine a few actual precursors. We must illustrate how the generation or so preceding 1660 wrote about nondemonstrative evidence. We shall show that the arcane events I have described, taking Paracelsus as model, have become encoded as the commonsense of the time. Sign-as-evidence has become a fixed point on the intellectual scene, not a matter for debate or reflection. I argue this not by 'interpreting' the words of early seventeenth century thought, but by quotation of the actual sentences that had become current.

First, the metaphor of the 'Author of the Universe' became endemic. Even Galileo could find it convenient to talk that way, and the lesser lights of the time did so everywhere. Here is Galileo, deliberately popularizing and using the commonsense of the time to argue for a more mathematical approach to physics:

Philosophy is written in this grand book, the universe, which stands continually open to our gaze. But the book cannot be understood unless one first learns to comprehend the language and read the letters in which it is composed. It is written in the language of mathematics [1957, p. 237].

That sort of talk was everywhere. Mention of signs and probability was not quite so universal, and so we can observe, in particular cases, the groping for a conceptualization which was achieved only around 1660. Here let us take as examples only the three great philosophers of the time, Descartes, Gassendi and Hobbes. The first has no truck with the nascent concept of probability, but the other two seek it out.

Probability had no place in the schematism of Descartes. Although he had grave qualms about *scientia*, he still sought to demonstrate not only the laws of motion of the planets and the laws of refraction of light, but also that the blood must be red. Descartes did employ what many a modern philosopher calls induction: he argued from observed effects to hypothetical causes. But he insisted that even though no scholastic would call that 'demonstration', it still was, in common speech, called demonstration. He did think that the hypotheses he demonstrated were mere fictions. Historians usually say this is because he was afraid of the sort of persecution that fell on Galileo, but there are deeper reasons. That was the only way in which he could fit the new hypotheses into the old theory of demonstration. In the waning distinction between high and low science, Descartes firmly opted for the high, and thereby determined the course of his philosophy. It had no room for probability.

As a fulcrum between Gassendi and Descartes we may usefully consider Herbert of Cherbury. His book *De veritate* was published in France in 1624. Descartes liked it and said so to his friends. Gassendi wrote a tract against it. After a theory of knowledge Herbert presents, in successive chapters, a sliding scale: truth, revelation, verisimilitude, possibility and falsehood. The 'verisimilitude' is entirely based on testimony. Herbert had only the old theory of *opinio* derived from witnesses or authority. That was fine for Descartes who thought that demonstrative science was possible. It was anathema to Gassendi, who could contend, *Quod nulla sit Scientia, et maxime Aristotelea*. There is *no* demonstrative science! This is the heading of a celebrated section in his book attacking the Aristotelians [1658, *Lib.* II, *Ex.* vi].

We use the expressions 'to have an opinion' and 'to know' interchangeably, as the practice of everyday speech shows, and if you look at the matter carefully, knowledge and opinion can be considered synonyms [1658, II, vi, 6].

Here is a nice example of the impotence of linguistic philosophy. Descartes said we commonly use 'demonstration' for inference of hypothesis from effect, so *scientia* stands inviolate. Gassendi said science and opinion are synonyms, and thereby denounced the old interpretation of demonstration. Descartes and Gassendi were both apostles of the new science, but they were pulling it in opposite ways.

Gassendi is first in the great line of empiricist philosophers that gradually came to dominate European thought. Unlike Francis Bacon (to whom this accolade is usually given) he did not try to revise the theory of *scientia* but to demolish it. He was sufficiently a scientist that he did not risk scorn by trying to work out the methodology of the empirics, but rather sought for ancient models. Much of his laborious scholarly production is directed at just this end. He did not find what he wanted until late in the 1630s. He wrote several chapters of the *Syntagma* about 1636, in which *signa* play no serious theoretical role. After he had devoted serious study to Sextus Empiricus, 'signs' are everywhere. [Bloch 1971, p. 145]. He had learned of the stoic conception of *signa*, which are either indicative or what Gassendi translates from the Greek *hypomnestika* as 'probable'. Modern translators prefer to call the latter 'associative' or 'suggestive'. Smoke is an associative sign of fire because smoke and fire have often been observed together, so that the presence of smoke calls fire to mind. O. R. Bloch summarizes the consequences of Gassendi's use of this stoic doctrine:

> It is in terms of signs that Gassendi developed his account of all kinds of scientific reasoning, accumulating astronomical and geometrical examples in order to show that it is through the use of signs that the astronomer and the mathematician become able to establish hidden truths [*Ibid.*, p. 146].

According to Gassendi, even syllogistic proof is a matter of signs, for the middle term in a syllogism is a sign. There was many a Pyrrhonist of the day who could echo Gassendi and say that there is no science. But Gassendi did not take the next step to total scepticism. On the contrary, the old demonstrations are preserved by his theory of indicative signs, and less certain knowledge is analysed by the theory of probable signs.

A great deal has, however, happened to the concept of sign when it passed from the language of the physician to the sign which is the deliberate, conscious, and understood expression of internal evidence. It is necessary, for example, to make the distinction between natural signs, and conventional ones. Paracelsus, remember, classed words with comets, halos, and statues. He thought that the (true) names of the stars are signs in exactly the way in which the points on a stag's antlers signify the animal's age. Of course it had always been realized that we can choose names at will, but wilful names were not true signs at all. The physician, chemist and astronomer must aim first and foremost at discovering the correct names of things. There is no element of convention in that. The discovery that all names are conventional thunders us into modern philosophy.

Arbitrary and conventional signs are carefully distinguished in the Port Royal *Logic*, the same book from which I took my terminology of internal and external evidence. Hobbes also very sharply distinguishes 'arbitrary' and 'natural' signs. Once natural signs have been distinguished from any sign of language, the concept of internal evidence is also distinguished. Hobbes was also able to record, almost casually, the connection between natural signs and the frequency of their correctness. By 1640 he wrote:

This taking of signs by *experience*, is that wherein men do ordinarily think, the difference stands between man and man in *wisdom*, by which they commonly understand a man's whole ability or *power cognitive*; but this is an *error*; for the signs are but *conjectural*; and according as they have often or seldom failed, so their *assurance* is more or less; but *never full* and *evident:* for though a man have always seen the day and night to follow one another hitherto, yet can he not hence conclude they shall do so, or that they have done so eternally: *experience concludeth nothing universally.* If the signs hit twenty times for one missing, a man may lay a wager of twenty to one of the event; but may not conclude it for a truth [*Human Nature*, IV. 10].

Here, in a text published in 1650, probability has emerged in all but name. The new internal evidence and the witting cognizance of frequency go hand in hand in a glove that bears one word: 'sign'. The space, in which the concept of probability was to emerge, is complete.

6

THE FIRST CALCULATIONS

The first faltering steps towards a European arithmetic of games of chance have been well chronicled by Øystein Ore [1953, 1960] and others. The only sixteenth-century book on the subject was not published until 1663, but throughout the sixteenth century various exercises on random phenomena occur in the commercial tracts that we now recognize as the start of European algebra. There are two fairly distinct aspects of these anticipations of probability theory: combinatorial problems, and problems about repeated gaming. The latter concern the division of spoils in an incompleted game. They form part of a large corpus of 'division problems' that arise in trade, and most of which have no aleatory basis. No one was able to solve those that were related to chance. It would be a brave interpreter of these early exercises who could assert, with confidence, that the authors had any idea that they were dealing with a new kind of subject matter. They were attacking one of a host of problems of 'fairness' that beset the new merchant class, and probability, in general, had nothing to do with these.

Combinatorial problems have a different kind of history. They become hooked up with the division problem in a clear and recognizable way only in the time of Pascal. Roughly speaking, combinatorial problems remain thoroughly in league with the alchemical magic of signs until the sign itself is liberated from that background in the seventeenth century. The great alchemist Raymond Lulle (1234–1315) is usually cited as the founder of the theory of combinations. He hoped to represent all the elements of the world by their true signs and then, by generating all possible combinations of signs, to produce true signs for all possible compounds in the universe. One would then know how to make any possible thing. The great combinatorial work follows this tradition. In a recent article Eberhard Knobloch [1971] shows how this

49

motivation persists quite explicitly even in Leibniz's *Ars combinatoria*, which is probably the first monograph on the theory of combinations, and which Leibniz himself came to see was integrally related to probability theory. Although the spectre of the old theory of signs long infected combinatorial arithmetic, I have discussed that enough, and in this chapter we can attend more to the actual calculations, whatever their occult origins.

We cannot entirely exclude astrology, magic, and signs even if we restrict ourselves to dicing. The first known enumeration of combinations of outcomes from three dice, described in Kendall [1956], is part of an explicit device for reading the future. Each throw has associated with it a particular concatenation of astrological elements, so that the heavens are divided into 216 units corresponding to the possible outcomes with three dice. Apparently one threw the dice to obtain a cheap and unreliable horoscope. No magus can have had much truck with such matters. I have argued that the conception of frequency and of partial degree of belief arise from the low sciences, but that transformation occurs at a level quite different from that of the charlatans with the dice.

We do find, in fifteenth-century Italy, genuine attempts to apply the new algebra to problems of gaming. The division problem is a touchstone throughout the whole period. Two (or several) players compete for a prize that is awarded after one of them has won *n* games. Player *A* has won more games than *B*, and through some intervention they must quit before reaching *n*. How should the prize be divided? Everyone agrees that *A* should get more than *B*, but how much more? In 1494 Luca Pacioli cast the problem in this form:

A team plays ball in such a way that a total of 60 points is required to win the game, and each goal counts 10 points. The stakes are 10 ducats. By some incident they cannot finish the game, and one side has 50 points and the other 20. One wants to know what share of the prize money belongs to either side. [1494, fol. 197 r.]

Ore [1960] says he has found this kind of problem in an Italian manuscript of 1380, and conjectures that the problem is of arabic origin. But people could not solve this problem: 'I have found that opinions about the solution differ from one man to another', Pacioli tells us, 'but all seem insufficient in their arguments. I state the truth and give the correct way to solve this problem.'

50

The first calculations

Cardano (1501–76), writing perhaps in 1564, says 'there is an evident error in the determination of the shares that even a child should recognize', but does not give a correct solution. In a similar case, with 6 goals, A having scored 5 and B 3, his rival Tartaglia (1499–1559) divided the prize 2:1. He seems to have reasoned as follows. A is 2 games ahead of B. This is $\frac{1}{3}$ of the total number of games required to win. Hence A should take $\frac{1}{3}$ of the stakes. The remainder is divided evenly, giving A an advantage over B in the ratio 2:1. Tartaglia himself was not optimistic about this solution, owning that 'the resolution of such a question must be judicial rather than mathematical, so that in whatever way the division is made there will be cause for litigation' [1556, I. fol. 266 1].

In 1558 G. F. Peverone did better. Let it be given that A will win if he takes the next game, and that he is staking one unit. If B also has only one game to go, he will stake one unit too. That is, the prize ought to be divided equally. If B has two games to go, he ought to pay 2 more units to get to the position of having only one game to go. Hence the prize should be divided as 3:1. If B has three games to go he ought to pay twice as much again, namely 4 units, to get to the position of only two games to go, and so the solution to Pacioli's problem would be 7:1. This at least seems to be his reasoning. 7:1 is also what we call the right answer, although by an apparent slip, contrary to his own rules, Peverone gives the answer 6:1. M. G. Kendall [1956] calls this one of the great near misses of probability mathematics.

Not all combinatorial problems were unsolvable. Galileo furnishes the most lucid examples of success. How many equal alternatives arise in tossing three dice? It was more common to play with three dice rather than two, and three dice were regularly used in telling fortunes. Kendall describes an accurate partition of outcomes that perhaps dates from around 1200. But in applying such partitions to probability we must exercise some care. The outcome 3, for example, can arise in just one way: 1–1–1. The outcome 4 also has one partition: 1–1–2. Unlike 3, the partition of 4 has three permutations: 1–1–2, 1–2–1, and 2–1–1. It is not obvious that permutations rather than partitions are equally probable. Indeed it is arguably an empirical fact about dice that can only be learned from observation.

In case the probabilities of compound outcomes with dice should now be so well established as to seem *a priori*, it may be useful to update the example. Consider elementary particles of microphysics. The natural generalization of dice takes r dice with n faces, thus giving n^r equiprobable alternatives – or, to express it differently, this is the distribution of r objects into n cells. This is called the *Maxwell–Boltzmann* system. So far as is known, no particles obey the Maxwell–Boltzmann laws, which apply to indistinguishable particles. If however we consider that *arrangements* of particles are indistinguishable, we get the *Bose–Einstein* statistics. If we add the further condition that it is impossible for two or more particles to be in the same cell, and assign equal probability to all arrangements satisfying that condition we get the *Fermi–Dirac* statistics. It is an empirical fact that photons obey the Bose–Einstein statistics, electrons obey the Fermi–Dirac statistics, and dice obey the Maxwell–Boltzmann statistics.

Leibniz once made the mistake of supposing that with dice partitions (rather than permutations) form the Fundamental Probability Set of equal alternatives. That is, he accepted the Bose–Einstein statistics for dice! Galileo was less prone to error. In a brief memorandum he relates that someone has been puzzled by a seeming contradiction between two facts. With three dice '9 and 12 can be made up in as many ways as 10 and 11'. Each, that is, can be decomposed into 6 partitions. However 'it is known from long observation that dice players consider 10 and 11 to be more advantageous than 9 and 12'. Galileo's solution is immediate. There is a 'very simple explanation, namely that some numbers are more easily and more frequently made than others, which depends on their being able to be made up with more variety of numbers'. In particular the 6 partitions of 9 and 12 break down into 25 permutations, while the 6 partitions of 10 and 11 decompose into 27 permutations. If permutations are equally probable, then 11 is more advantageous than 12 in the ratio 27:25.

It seems natural to call this our first witness of the refutation of a statistical hypothesis by 'long observation'. We have the hypothesis that permutations are equally probable versus the hypothesis that partitions are equally probable; the latter is inconsistent with the facts, while the former fits the facts perfectly. It would have been easier to test by observing the relative proportion of 4 and 3, of which there are the same number of partitions but 4 has three times as many permutations. However, in standard dice games 4 and 3 were 'hazards', that is to say, numbers deemed too difficult to make, and such that all bets were called off when 3 or 4 occurred. Hence long experience was not recorded, whereas the fortunes of those who bet on 9 through 12 were well known.

I have described Galileo's problem as a conflict between experimental result and a particular theory. Maistrov [1974] disagrees. He thinks that the ratio 27:25 is too near equality for anyone to notice the discrepancy. There was, after all, no statistical technique for hypothesis testing, and Italian dice were probably imperfectly symmetric, adding distortions that would outweigh any tendency to get 12 slightly more often than 11.

This problem of interpretation recurs, especially in connection with the Pascal–Fermat correspondence. Are we to think of these probability puzzles as arising in part from empirical data, or are they of purely arithmetical origin? It is impossible to answer with certainty. On the one hand Galileo actually tells us that 'long observation' produces the puzzle. On the other hand, we know that taking averages of results was rather foreign to science until Galileo himself started taking means. Yet we may throw in the balance an observation of Kendall's. Long before anyone could prove, mathematically, that in poker a straight is more common than a flush, it was a standard part of the game to value the flush more highly. Indeed in general the ranking and values of various tosses was known with considerable exactitude. Kendall cites craps according to whose rules the chance of the first player ought in fairness to be $\frac{1}{2}$. In fact it is very close, 244/493. Thus there exists a great deal of correct empirical lore. Hence I am inclined to believe exactly what Galileo reports, that 'long observation' had taught people that 11 is better than 12.

Galileo had a sense of probability concepts worthy of his genius. Maistrov [1974] and Eisenhart [1971] dwell on his sophisticated appreciation of the method of combining discrepant observations, while Todhunter [1865, p. 5] notes a curious controversy on the relative virtues of the arithmetic versus the geometric mean. One feature, however, is not peculiar to Galileo. He is explicitly concerned with the relative frequencies of different outcomes. At this stage in history the probability calculus is a calculus of frequencies. These evident facts need emphasizing because Carnap said that in 1866 John Venn 'was the first to advocate the frequency concept [. . .] unambiguously and systematically' [1950, p. 186]. In case this suggests that the frequency concept (as opposed to a subjective or logical concept of probability) emerged only in 1866, we must insist on its unambiguous presence three centuries earlier. No one can doubt this after reading the first book about games of chance. It was written by Cardano around 1550 – the date 1564 is often canvassed. It was not printed until 1663, but Cardano's lectures were popular and although his work could not have much direct influence on subsequent generations, he does serve to show what people of his time thought was their object of enquiry. Here is a typical remark:

> In three casts of two dice the number of times that at least one ace will turn up three times in a row falls far short of the whole circuit, but its turning up twice differs from equality by about $\frac{1}{12}$. *The argument is based upon the fact that such a succession is in conformity with a series of trials and would be inaccurate apart from such a series* [*De ludo aleae*, Ch. 11].

We wonder exactly what Cardano meant by 'the whole circuit'. It is something like the number of equiprobable outcomes. The rest of this quotation is plain enough. Venn, the alleged first frequentist, wrote that 'the fundamental conception which the reader has to fix in his mind as clearly as possible, is, I take it, that of a series. But it is a series of a peculiar kind, one of which no better compendious description can be given than that which is given by the statement that it combines individual irregularity with aggregate regularity' [1866, p. 4]. We have no doubt that Cardano had fixed this idea in his mind. Another characteristic remark of Cardano's is instructive:

> I am as able to throw 1, 3 or 5 as 2, 4 or 6. The wagers are therefore laid in accordance with this equality if the die is honest, and if not, they are made so much the larger or smaller in proportion to the departure from true equality [Ch. 9].

David [1962, p. 58] singled out this passage as the real novelty in Cardano's work. For the first time, she says, we find an explicit idealization to a number of equal alternatives based on the abstraction of a fair die. Philosophers will find additional interest in such passages. There has recently been much discussion of a 'propensity' interpretation of probability by Karl Popper which is known to have been anticipated by C. S. Peirce. The idea is that a probability statement about dice, say, asserts a propensity, tendency, or disposition of the dice to display certain stable frequencies on repeated trials. There is some infighting as to whether such propensities can be sensibly applied to single cases or not, but we evade that issue here. It is clear that Galileo's phrase about some outcomes

The first calculations

being 'more easily or frequently made' refers to the propensities of the dice. Cardano, in the above quotation said 'Tam possum proiicere [. . .]' speaking of his 'ability' to throw various combinations. Indeed we shall see that such dispositional terms are subsequently supplemented by talk of the proclivities of the dice to yield various outcomes. 'Proclivity' in the dictionary is a synonym for propensity! The word 'probability' is not used in these connections for another century. In this respect one must be cautious of translations. For example Gould, translating Cardano, has: 'the desired result may or may not happen with equal probability' [Ch. 11]. Cardano had written, *in totidem enim potest contingere & non contingere*. If we seek a smooth translation, that of Gould is fine. But it does not imply that the probability of opinion, described in Chapter 3 above has been formally expressed in the theory of chances.

There is, however, no anachronism in taking such expressions as Cardano's to show that he had some sort of 'propensity' attitude to aleatory chance. That still leaves open how Cardano conceived of the equal propensity, proclivity, ability or facility of a die to land on each of its six faces. We have very little comprehension of how the Renaissance conceived any sort of influence, power, or potentiality. As well as being a gambler and an algebrist, Cardano was a physician and a metaphysician. His fame as a doctor of medicine ran from Italy through Paris and the Low Countries even to Scotland, where he was brought, at great expense, by the Duke of Hamilton. We ought to obtain some notion of his concept of causality from his ample writing on nature and on disease. At present that can only be a project and a programme.

Michel Foucault has used Cardano in counterpoint to Paracelsus. Where Paracelsus preached the doctrine of similitude, Cardano hewed to the older line of antipathy [1970, Ch. 2]. But Cardanesque antagonism and Paracelsian similarity constitute theories born from the same intellectual womb. In neither Paracelsus nor Cardano can we readily grasp what 'makes things happen'. Nor, I think, did such thinkers grasp it themselves. Like all their fellow empirics they could not divorce themselves from the medieval view of causality that was still professed by high science. Yet there was no place in their day-to-day work for causes that could be proven from common notions and first principles. Cardano seems to have tackled the problem seriously in his book, written in 1550, on 'subtlety'.

55

Subtiltas, we learn from the first page, is almost the very medium of causality, or at least, the medium of interaction of all things. It is that through which antipathies have their effects, it is that through which the mind takes in data from the senses and renders them intellectual. It is that through which the heavenly bodies determine the destiny of man. It is that through which nature reveals itself to us in signs.

It may be admissible to see Cardano and other students of chance adopting a 'propensity' attitude towards dice: they conceive the dice as having various tendencies to fall in various ways. But it is quite improper to identify this concept of 'tendency' with anything that has recently gained a place in modern philosophy. Propensities have been introduced by current thinkers to place random phenomena in the confines of twentieth century concepts of causality. When the concept of cause in the Renaissance is so alien to ours, and is indeed being transformed, unwittingly, by the very writers whom we study, it would be absurd to identify any Renaissance metaphysics of chance with any that is current today.

What of the claim that probability emerged only in 1660? Do we not find all the germs of a reflective study of chance in Cardano? Yes indeed: it has been no part of my thesis that there were never precursors nor anticipations. Cardano was physician extraordinary, proud astrologer, and mathematician enough to try to claim that he had solved the cubic equation. He was deeply aware, in his own way, of many of the ingredients that in due course coalesced to form the space in which probability did emerge. It is not so noteworthy that there was such a precursor, but rather, that although he was an eminent if bizarre character, his work on chance sank into oblivion. The book was not even published until the birth-decade of probability. When the division problem and the like were again taken up just before 1660, their Italian sources were entirely forgotten. In none of its forms, and by none of its names, had the new probability entered the discourse of Europe until that time. I do not in any way mean to play down the interest of that historical figure, Cardano. Our study is not of great men but rather of an autonomous concept. Some may be bold enough to assert that out of some historical necessity probability could not have become public property at the time of Cardano. More modestly we say only that it did not. It waited another century.

7

THE ROANNEZ CIRCLE

Poisson's famous sentence puts the matter succinctly if inaccurately: *A problem about games of chance proposed to an austere Jansenist by a man of the world was the origin of the calculus of probabilities* [1837, p. 1]. Leibniz reports it at greater length: 'Chevalier de Méré, whose *Agréments* and other works have been published – a man of penetrating mind who was both a gambler and philosopher – gave the mathematicians a timely opening by putting some questions about betting in order to find out how much a stake in a game would be worth, if the game were interrupted at a given stage in the proceedings. He got his friend Pascal to look into these things. The problem became well known and led Huygens to write his monograph *De Aleae*. Other learned men took up the subject. Some original in early fixed Phenomena de Witt used them in a little book on annuities printed in Dutch' [*P.S.* v, p. 447].

This story is so familiar that it is, perhaps, the only incident in the early history of probability that may be called common knowledge. Since the circumstances have aroused such general interest, I shall depart from my usual practice and say a little more about what actually happened. When we pass from a study of the texts to more anecdotal concerns, we shall in general find that Leibniz, whom I have just quoted, is an admirable witness. He made no serious formal contribution to probability theory, but he had a lasting and profound interest in the subject. Indeed he was the first philosopher of probability. He was the first to insist that probability theory can serve in a branch of logic comparable to the theory of deduction. Before coming to Paris, and before mastering the work of 'Pascal, Huygens and others' he had tried to develop an arithmetic of probability that was not based on games of chance, and hence was potentially more general in application. He also wrote the first monograph on the theory of combinations, and saw its relation to probability theory. He was the first to try to axiomatize probability as a pure inferential science. He saw how a generalized theory of games should be the foundation for making any quantitative decision in situations where one must act on inconclusive evidence. If we abstract him from his environment, he often seems a twentieth century figure. So it is useful, to some extent, to watch the growth of probability arithmetic through his eyes. He knew all the protagonists except Pascal, who

had recently died when I lived in Paris, but his sister was there, and also his nephews, the sons of his sister. I had many associations with them, as well as with the Duke of Roannez, who had been a close associate of Pascal's, and I was much influenced by these studies. From them I received some of Pascal's unpublished works to read, though they were mostly mathematical [*S.S.* II. I. p. 533].

Roannez was a gifted amateur whose salon was the meeting place of the mathematicians and the gentlemen of Paris. If Poisson were right, we could call Roannez the midwife in the birth of the probability calculus. The social events surrounding that occasion have been amply reported, for the relation between the 'austere Jansenist' and the 'man of the world' has attracted the imagination of generations of scholars. There is a letter in which Méré writes of a coach trip to Poitou with his friends the Duke of Roannez and Damien Mitton. The duke 'had a mathematical mind, and in order not to be bored on the way, he had brought along a man neither old nor young, then very little known, but who has since made a name for himself. This man was a great mathematician but at the time he knew nothing but mathematics.' Méré relates how this fellow gradually learned true wit from his fashionable fellow travellers and resolved to give up mathematics, all in a three day coach trip.

François Collet [1848] made the happy guess that this mathematician must have been Pascal. Here was the very occasion on which Méré presented his problem and probability theory began. Dramatically, too, we have a clue to Pascal's occasional renunciations of mathematics. Collet's conjecture fitted the known facts fairly well, but not so perfectly as to prevent Fortunat Strowski casting Descartes for the coach ride! [1930, II, p. 231]. Strowski was certainly mistaken, but was Collet right? Jean Mesnard, the celebrated authority on Pascal biography devoted 1100 pages to the relation between Pascal and Roannez [1965]. His judicious survey of the documents leads him to the tentative verdict, 'not proven'.

Then, after prodigious efforts to find three days during which Méré, Mitton, Roannez and Pascal could all have been in the same coach, he concludes that such a date is 'unfindable'. Méré could not after all have been writing about Pascal.

It does not matter. Roannez was a constant friend and patron of Pascal, and a companion of Méré. The Pascal–Fermat correspondence did arise from these friendships. According to Leibniz, it was Roannez who got Pascal to re-enter the learned world after he had withdrawn to Port Royal. Pascal had solved many problems on the cycloid, and Roannez proposed that Pascal should run a competition and win added acclaim as a mathematician. This could only improve the force of the apology for religion, which we now call the *Pensées*. This was particularly important because the apology was to include a quasi-mathematical argument for belief in God. The greater the fame of the author the more he would prove convincing.

The author's fame had, of course, been amply established already. No matter how problems about probability were first presented to Pascal, his solution of them became well known, for the solution to the division problem was printed at the end of Pascal's monograph on the arithmetic triangle. The first question which Méré asked Pascal is, in itself, a trifling one. In throwing two dice, how many tosses are needed to have at least an even chance of getting double six? Méré thinks two answers are possible, 24 and 25. It may look as if this is another conflict between theory and experiment similar to that which prompted Galileo's note on tosses with three dice. The answer 25, it may appear, is obtained by experience, while the answer 24 is got by a mistaken arithmetical rule. Ore argues powerfully to the contrary [1960]. To begin with, the discrepancy between the probabilities for 25 and 24 is small – 491 : 505. The minuteness of this difference does not in itself prove that it could not have been noticed in the course of ample experience of gambling. But Pascal tells us that Méré thought his problem showed that arithmetic contradicted itself. This suggests that it is a dispute among mathematical theories that prompts the problem, and not a conflict between experience and theory. Then Ore actually reconstructs the alleged contradiction which was intended to show that arithmetic totters (or at least, *se démentoit*).

Pascal had little difficulty with Méré's 'great scandal'. An elementary enumeration of possibilities shows that in tossing two dice the probability of getting double six in 24 throws is only 0.491, while in

25 throws, it is 0.505. Indeed Pascal has some difficulty understanding how Méré could get any other answer. Ore has explained Méré's reasoning. A rule old enough to be found in Cardano runs as follows. Consider a set-up in which we have one chance in N of winning at a single trial. Let n be the number of trials required to have a better than even chance of success. Méré's rule, apparently, is that n/N is constant. In the case of a single die with which we try to make an ace, say, N is 6 and n is 4 (because after 4 trials we have 671/1296 chances of getting at least one ace). So n/N is 2/3. Hence if N is 36, as for two dice in trials aiming at double six, n must be 24. This is the source of the incorrect answer 24. As it happens, Méré's curious rule is not absurd, but, on the contrary, is asymptotically correct. For large numbers it may even have been experimentally well confirmed. Hence there is a sense in which Méré's problem arises from a conflict between experiment and theory. The correct resolution is to note that the rule is only asymptotically sound, and does not work for small N.

The general solution to the division problem, on the other hand, was quite something, for it produced the first thorough understanding of binomial coefficients by way of the arithmetical triangle. That triangle of numbers had been devised before, by Cardano, for example, and by the Chinese, but no one well understood what it meant. Indeed, arguably this was the only significant mathematical discovery (as opposed to conceptual discovery) in probability until at the end of the century Jacques Bernoulli discovered limiting properties of these same coefficients.

As regards the immediate question, the division problem, Pascal made a step which looks odd. We are concerned to divide the stakes when one player needs n games to win and the other needs m. Pascal enumerates all cases in which $n + m - 1$ games are played, although commonly the contest would come to an end before that many games were completed. Pascal wrote to Fermat asking for confirmation of his solution, which Fermat was glad to give. These letters were soon in the public domain. Poisson is perfectly right to attribute the foundation of the probability calculus to these letters, not because the problems were new, nor because other problems had not been solved by earlier generations, but because here we have a completely new standard of excellence for probability calculations.

Problems in probability arithmetic were amply, though often not very seriously, discussed. Roannez set several for Leibniz to solve. Some are comparative trifles about dicing, but a question about annuities is of some importance, as we shall see. Leibniz is also partly responsible for the perpetuation of the story about the relation between Méré and Pascal. Méré is in fact not mentioned by name (other than by his initial) in the Pascal correspondence. In the 1676 notes on the Roannez problems, Leibniz refers to 'Mesle, Pascal and Huguens' (sic). On several occasions he recites the story I quote at the beginning of this chapter. Oddly he seems to have forgotten it later. At any rate he asked another Roannez visitor – des Billettes – for confirmation of this history and got it in 1696. In the 1702 edition of Bayle's dictionary he attributed the Méré–Pascal story to Billettes. It is in that same dictionary that a curious and inflated letter from Méré to Pascal is published – 'You know that I have discovered such rare things in mathematics that the most learned among the ancients have never discussed them and they have surprised the best mathematicians in Europe. You have written on my inventions, as well as Huygens, Fermat and many others who have admired them.' Leibniz's reaction is characteristic: 'I almost laughed at the airs which the Chevalier de Méré takes on in his letter to Pascal.'

Méré was for a long time regarded as the 'inventor of division in games', which is how Huygens describes him, recounting a dinner party at the Roannez. Huygens' book on games of chance is the first printed textbook of probability. He had been working on problems of probability theory about the time of the Pascal correspondence, and tried to meet Pascal in 1655, without success. So he went back to Holland, wrote his own problems, and had them circulated. Fermat in particular approved, and sent two problems, appended to Huygens' book. Moreover, although Pascal was on retreat at Port Royal, he did convey to Huygens a gambling problem also included at the end of Huygens' little book. Nor was personal contact finished when Pascal went into seclusion. We know from Huygens' splendid diary of a second visit to Paris, in 1660, that he saw something of Pascal, and was in fairly regular conversation with other members of the Roannez circle. Probability is only a minor aspect of the intellectual ferment within that group, which has been amply studied elsewhere.

Pascal is properly remembered as the first significant figure in probability theory. But this is, I think, partly for the wrong reason.

The correspondence with Fermat was intrinsically important, and it got Huygens to work on the subject. But there is a quite different, and more generally important, contribution that Pascal made to the appreciation of the new concept of probability. This has never been taken seriously by historians of probability, and has chiefly been the preserve of apologists of religion. It is, however, the first contribution to what we now call decision theory, and, as I shall show, a very thorough one.

8

THE GREAT DECISION

Pascal's wager (*le pari de Pascal*) is the name given to some game-theoretic considerations that concern belief in God. They were intended as a contribution to apologetics, and became very widely known as such. But these fragments in the *Pensées* had an important byproduct: they showed how aleatory arithmetic could be part of a general 'art of conjecturing'. They made it possible to understand that the structure of reasoning about games of chance can be transferred to inference that is not founded on any chance set-up.

The wager is not easy to understand. Logicians have dismissed it. They have been mistaken. Pascal's pages contain three distinct arguments. Each is valid. Each has the form of a decision theoretic argument of a sort properly classified and characterized only in this century. Although Pascal did not state his underlying principles, it seems clear that he did know what he was doing. The reasoning was novel, but the popularity of the *Pensées* made it a familiar fact that games of chance could serve as models for other problems about form of decision under uncertainty.

The wager occurs in a passage headed *infini – rien*, no. 418 in Lafuma's numeration, and 233 in Brunschvicg's. It consists of two pieces of paper covered on both sides by handwriting going in all directions. It is full of erasures, corrections, and seeming after-thoughts. For a photograph, see Brunet [1956]. There is endless speculation about when and how Pascal wrote these two pages. Every blot of ink or raindrops on them has been given minute analysis, for although logicians have so often denounced the argument it has had a striking attraction for many religious minds. This has been especially true in our own times when the existential and non-deductive features of the argument have seemed more important to moralists and theologians than traditional, more deductive, ways to God. Despite this immense amount of scholarship most scholars have been aware only of the rough lines of Pascal's thought, and have not discerned its fine structure. So it is wise to begin with a modern statement of the arguments in question, and then show how Pascal actually deploys them.

Decision theory is the theory of deciding what to do when it is uncertain what will happen. Given an exhaustive list of possible hypotheses about the way the world is, the observations or experimental data relevant to these hypotheses, together with an inventory of possible decisions, and the various utilities of making these decisions in various possible states of the world: determine the best decision.

A special case of this problem occurs when no experiments are made. In the thought that concerns us, Pascal deliberately 'ties his hands' and refuses to look at any observations or experimental data bearing on the existence of a Christian God. He is writing for the man who will not countenance miracles, or the doctors of the church, or the witness of the faithful. So we may restrict our attention to the logic of decision when there are no experimental data.

Among the valid argument forms investigated by decision theory, Pascal apprehended three. I call them 'valid' in the sense now favoured by logicians. A valid argument form is one in which the conclusion follows from the premises. Colloquially, of course, a valid argument is one that is pertinent and persuasive. Pascal's arguments will not now be found persuasive. This is not because they are invalid (in the logician's sense) but because the premises of the arguments are, at best, debatable. Here are the three argument forms that concern us:

(*a*) *Dominance*. The simplest special case occurs when one course of action is better no matter what the world is like. Schematically, suppose that we have some exhaustive set of possible states of affairs: we label the states $S_1, S_2, \ldots$. Suppose that in some state S_i the utility U_{i1} of performing act A_1 is greater than the utility U_{ij} of performing any other act A_j. In no other state of affairs is the utility of performing A_1 less than A_j. Then, under no circumstances could A_j have happier consequences than A_1, and under some circumstances A_1 could be better than A_j. A_1 is said to *dominate* A_j. If some acts dominate all others the solution to our decision problem is, perform *a dominating act*.

(*b*) *Expectation*. The argument from dominance does not con-

64

sider how likely are various states of affairs. Even if dominating A_1 is better only in a very unlikely state of affairs, then, since A_1 can never fare worse than any other act, it is worthwhile performing A_1. But suppose no act dominates, although we do think we know which states of affairs are more likely than others. Suppose we can assess the probability of each state of affairs. Then (no matter what one means by 'probability') one argues as follows. We have assigned a probability p_i to each possible state of affairs S_i in some exhaustive set. Let U_{ij} stand for the utility of doing A_j if S_i actually obtains. The expected value, or expectation, of A_j is the average value of doing A_j: namely $\sum_i p_i U_{ij}$. An argument from expectation concludes with the advice: Perform an act with highest expectation.

(c) *Dominating expectation.* It may happen that we do not know, or cannot agree on, the probabilities of the various states of affairs. We can at best agree on a set of probability assignments to the states S_i. For example, suppose we agree that the coin is biased toward heads, but disagree how great is the bias; at least we agree that the probability of 'next toss gives heads' exceeds $\frac{1}{2}$. If in some admissible probability assignment, the expectation of A_1 exceeds that of any other act, while in no admissible assignment is the expectation of A_1 less than that of any other act, then A_1 has dominating expectation. The argument from dominating expectation concludes: Perform an act of dominating expectation.

The three argument schemes are mutually consistent. If one act does dominate the rest, then it will be recommended by all three arguments. If there is no dominating act, but there is an act of highest expectation, that act will also be the act of dominating expectation. The argument from dominance is the rarest, most special case. The argument from dominating expectation is more widely applicable.

Pascal's procedure in the thought *infini – rien* is to offer an argument from dominance. But if this is rejected, another premise is added and we obtain an argument from expectation. Then, if the second lot of premises be rejected, he offers an argument from dominating expectation. Not all the links forged by Pascal are transparent, but it is remarkable how these two scratched sheets that constitute *infini – rien* conform to this analysis. Here are the details.

The argument is directed at the sort of person who, not being convinced of the proofs of religion, and still less by the arguments of atheists, remains suspended between a state of faith and one of unbelief. This assertion is extremely important. A decision problem requires an exhaustive partition of possibilities. It is taken as a premise of the argument that *either* there is no God, *or else* there is a God whose characteristics are correctly reported by the Church. The God of the Muslims, for example, is not admitted as a possibility. It is a corollary that Pascal's argument is good for any decision problem with the same formal structure. 'An Imam could reason just as well this way', as Diderot remarked [1746, 'Addition' LIX]. Pascal's partition of states of affairs may be out of place today, but this is one thought from a book of thoughts. The other thoughts contain other reasons bearing on this partition. There are also other arguments directed at other special sorts of person, for example, the arguments directed at Orthodox Jews whose partition, of course, is not at all like that of the Parisian man about town.

'God is, or he is not' is Pascal's expression of his partition. 'Which way should we incline? Reason cannot answer.' That is, there is no valid proof or disproof of God's existence. Instead we adopt the following model:

A game is on at the other end of an infinite distance, and heads or tails is going to turn up. Which way will you bet?

The model is then reinforced. When reason cannot answer, a sensible man can say that he will not play the game. But in our case, by the mere fact of living, we are engaged in play. We either believe in God, or we do not.

In addition to a partition of states of affairs a decision problem requires a list of possible actions. As Pascal sees it, you either act with complete indifference to God, or you act in such a way that you will, in due course, believe in his existence and his edicts. There is no cant to Pascal. He accepts as a piece of human nature that belief is catching: if you go along with pious people, give up bad habits, follow a life of 'holy water and sacraments' intended to 'stupefy one' into belief, you will become a believer. Pascal is speaking to a man unsure whether to follow this path, or whether to be indifferent to the morality of the church. The two possible acts are not, 'Believe in God' and 'Do not believe.' One cannot decide to believe in God. One can decide to act so that one will very probably come to believe in God. Pascal calls that the wager that God is. To wager that He is not is to stop bothering about such things.

The great decision

The decision problem is constituted by two possible states of the world, and two possible courses of action. If God is not, both courses of action are pretty much on a par. You will live your life and have no bad effects either way from supernatural intervention. But if God exists, then wagering that there is no God brings damnation. Wagering that God exists can bring salvation. Salvation is better than damnation. Hence the wager, 'God is', dominates the wager, 'He is not'. The decision problem is solved by the argument from dominance.

The argument is valid. The premises are dubious, if not patently false. Few non-believers now can suppose that Pascal's partition exhausts the possibilities. If we allow just one further alternative, namely the thesis of some fundamentalist sects, that Jehovah damns all who toy with 'holy water and sacraments', then the Catholic strategy no longer dominates. That is, there is one possible state of affairs, in which the Catholic strategy does not have the best pay-off.

It is a question for the historian whether Pascal's argument from dominance could ever have been effective. Perhaps some Parisians three centuries ago did believe, say, on the evidence of alleged revelation, that if there is any supernatural truth, it is that truth professed by the Church. Such people may have found the argument persuasive.

From the very first, a fallacious correction has sometimes been urged. M. J. Orcibal has found the following contemporary note by Daniel Huet: 'this reasoning suits all religions; that which proves too much proves nothing. It proves only the necessity of having some religion, but not the Christian religion' [1956, p. 183]. This is a mistake; unless one has independent grounds for excluding the proposition of an eccentric religion, that all and only religious people are damnable. This point has been nicely made recently by James Cargile [1966].

One premise needed for Pascal's argument is that faith is catching. His interlocutor does protest, that perhaps a treatment of holy water and sacraments is not going to work. The theme is not well developed. One may pedantically urge a partition into three states of affairs: God is not; God is and one will come to believe by being pious; and, God is but one will not come to believe by being pious. The pious strategy still dominates, under the initial assignment of utilities favored by Pascal. There are a good many other variations and extensions of the partitioning. Also, many of us will share William James' [1897] suspicion that a person who becomes a believer for the reasons urged by Pascal is not going to get the pay-offs he hopes for. Although all these questions are crucial to the merit of the argument as a piece of apologetics, they are irrelevant to its logical validity. The conclusion does follow from the premises.

So far we have stated only case (*a*), the argument from dominance. In the course of the argument we assumed that if God is not, then either course of action has roughly equal utility. But this is untrue. The libertine is giving up something if he chooses to adopt a pious form of life. He likes sin. If God is not, the worldly life is preferable to the cloistered one. The wager 'God is' does not dominate, for there is one circumstance (God is not) in which accepting this wager, i.e. adopting the pious life, has lower utility than the other wager. When dominance is challenged, we require an argument from expectation.

The transition to case (*b*) is swift, perhaps too swift. Having heard the argument from dominance, the interlocutor says, 'very well. But suppose I am asked to stake too much?' This remark is obscure unless seen in the light of the present analysis. The interlocutor is protesting that he has to give up something to follow the pious life. Perhaps he stakes too much, that is, perhaps there is a significant difference between the utility of the two actions, under the hypothesis that there is no God. So Pascal has to introduce the policy of maximizing expectation. If the chances of heads and tails are equal, and there are equal pay-offs for either outcome, it is a matter of indifference whether we bet on heads. But if heads pays twice as much as tails, then clearly we bet on heads. In the agnostic's existential situation, the optimal pay-off, if there is no God, is a worldly life. The optimal pay-off, if there is a God, is salvation, of incomparably greater value. Hence, if there is an equal chance of God's existence or non-existence, the expectation of choosing the pious life exceeds that of choosing the worldly one. The argument from expectation concludes: act so that you will come to believe in God.

This argument from expectation can hardly be maintained. Although it is valid, it is presented with a monstrous premise of equal chance. We have no good reason for picking $\frac{1}{2}$ as the chance of God's existence. This argument can work only for people who are, in the strongest sense, exactly as unsure whether God exists, as they are unsure whether he does not exist. Against all other agnostics, another argument is called for and so another premise is invoked.

The argument from expectation with an equal probability distribution requires only that salvation, if God is, is more valuable than sinful pleasures, when there is no God. But salvation is infinitely preferable to the joys of the worldly life. No matter how great may be the daily pleasures of the libertine, they are finite. Salvation, according to Pascal, is infinitely blessed. Moreover, although we have no idea of the chance that God exists, it is not zero. Otherwise there would be no problem. There is a finite positive chance that God exists. No matter what this finite chance is – no matter how small – the expectation of the pious strategy with infinite reward exceeds that of the worldly one. Hence the pious strategy must be followed. This is case (*c*), an argument from dominating expectation.

These three arguments are all valid. None are convincing. All rely on dubious premises. The arguments are worthless as apologetics today, for no present agnostic who understood the arguments would ever be moved to accept all the premises. The most dubious premise is the partition, with its concomitant assignment of utilities. God is (and belief in Him brings salvation), or, God is not (and non-believers who have heard the Word are damned). It is no criticism of Pascal that he assumes this partition: he is directing his argument at his fellows who accept it.

Many questions of morality and theology may be adduced against the wager. What is striking is that Pascal raised the logically relevant questions. There is a logically special feature of the argument from dominance, which is challenged if the utility of a libertine life exceeds that of a pious one, in the event that there is no God. To allow for this difficulty, we move to the second argument form, employing probabilities. An equal distribution of probabilities has some appeal, but that special logical feature is in turn abandoned as we move to the third argument form.

Throughout this analysis I have freely used the word 'probability'. It is not used by Pascal. All the technical terminology is that of the theory of gaming, of chances and hazards and coins. No one today would want to say that a chance set-up like a coin or loaded die determined whether or not God should exist. We would express the argument in terms of some idea of subjective or personal probability, saying, for example, that no matter how slender our degree of belief in the existence of God, it is not 0. Pascal does not speak of a quantitative measure of degree of belief. He is saying that we are in the same *epistemological* position as someone who is gambling about a coin whose aleatory properties are unknown. His judgement relies on an alleged isomorphism between the structure of a decision problem when objective physical chances are known to exist, and a decision problem in which there are no objective physical chances. It was Pascal's colleagues at Port Royal who first spoke of measuring something they actually called probability. Before I develop that part of the story, let us look briefly at how Pascal's wager was received by his immediate successors. Here our concern is not so much with the concept of probability, but with whether anyone actually understood what Pascal was doing.

The *Pensées* were not printed by the Port Royal editors until 1670, eight years after the death of Pascal. The immediate fame of the wager is not attributable only to this edition. The final page of the first edition of the Port Royal *Logic* [1662] has a brief summary of the ideas. Directly after the appearance of the *Pensées*, readers such as Leibniz discerned the source of those ideas [see a letter of 1678, *S.S.* I. II. p. 112]. Locke picked up Pascal's argument, like much else, from the Port Royal argument. (Indeed it has regularly been conjectured that Locke was one of the 'several hands' who translated one of the early English editions.) The actual wager is described in the *Essay Concerning Human Understanding* [II. 21. 72]. It excited a good deal of undeserved admiration. Locke thinks the whole argument follows for anyone who:

will allow exquisite and endless happiness to be but the consequence of a good life here, and the contrary state the possible reward of a bad one [. . .] I have forborne to mention anything of the certainty or probability of a future state, designing here to show the wrong judgement that any one must allow he makes, upon his own principles, laid how he pleases, who prefers the short pleasures of a vicious life upon any consideration, whilst he knows, and cannot but be certain, that a future life is at least possible.

Locke, it is evident, had no conception of probability logic. Leibniz, in his commentary on Locke called *Nouveaux Essaies*, simply passes by Locke's version of the argument. By the time he read Locke, the argument had no novelty. In a letter dated 1683, Leibniz's opinion on the merits of the wager is unequivocal. He grants that it may have some uses. Every man must put every argument to what use he can. However:

Pascal paid attention only to moral arguments, such as he presented so well in his little posthumous book of thoughts, but he did not put much value on the metaphysical arguments that Plato and St. Thomas and other philosophers and theologians have used in proving the divine existence and immortality of the soul. In this I do not agree with him. I think that God speaks to us, not merely in sacred and civil history, but also internally, within our mind, through truths which abstract from matter, and are eternal. Even if I should confess that these arguments have not been carried to the full force of a complete demonstration, they already seem to have as much force as the moral arguments. I believe that men will gradually perfect them. [A letter of 1683, *S.S.* II. I, p. 533.]

Pascal's opinion was exactly the opposite. 'The metaphysical proofs of God are so removed from the reasoning of mankind, and so complicated, that they have no force.' This thought, though appearing in all arrangements of the Pensées a long way from '*Infini-rien*', has been shown by Georges Brunet [1956 pp. 48–51] to be one of four jottings written on the same kind of paper as that Pascal used for his wager. It is inferred, with some likelihood, that it was much in his mind at the time.

Our long quotation may suggest that Leibniz was a little condescending about Pascal's style. Quite the contrary. A respected apologist named Michel Mauduit anonymously printed a version of the argument [1677]. Next year Leibniz contracts the merits of this book with that of Manuel Mauduit 'is not for people who reason with application, and who prefer naked thought as in Pascal, where it is stripped of useless words.' [*S.S.* I. II, p. 112.]

Mauduit is not that bad: he gives a studious explication of Pascal's arguments. He states, in the preface, that the discussion in Pascal 'is too short and too fast for something so important'. He fills it out, and treats objections. He states the simple principle of dominance as a maxim, 'used in commerce and all branches of life'. I have not found the maxim stated earlier. Despite such occasional insights, Mauduit's tract is singularly insubtle. He hardly realized that Pascal's version of the wager is based on the premise that faith is catching. Pascal is free from, but Mauduit deserves, these remarks of Leibniz:

This argument shows nothing about what one ought to believe, but only about how one ought to act. That is to say, it proves only that those who do not believe in God should act as if they did.

Pascal would add: because, if they do so, they will in due course come to believe. Now that is no reason for believing. It is a cause of believing. Leibniz is a rationalist about belief. Pascal understands belief better than Leibniz.

Travesties of Pascal's argument became common. In 1699 John Craig incorporated a version of it into his 'geometrical' arguments for faith. He had a theory about the testimony of witnesses. Certain events are said to have happened in the Holy Land a long time ago. How reliable are these reports? Craig supposes that the credibility of the stories diminishes at each telling. By now the stories have worn a little thin. But because they still have finite probability, and because of the infinite pay-offs for believing them if they are correct . . . and so on.

Craig's book was popular enough to be republished in 1755. In Chapter XI of his *Essaie philosophique* Laplace amused himself by demolishing the argument, and indeed turns it on its head. The credibility of a witness is in part a function of the story being reported. When the story claims to have infinite value, the temptation to lie for personal benefit is asymptotically infinite. Hence the biblical tales do not have even finite probability, and the argument collapses. Curiously, Laplace takes this to refute Pascal. But Pascal explicitly rejects any appeal to witnesses. Laplace can have read only Craig or something similar.

Whatever its value as an apologetic, Pascal's logic remained. Games were seen to serve as models of all sorts of decision under uncertainty. Voltaire was too late when in 1728 he said of *'Infini–rien'*,

This article seems a little indecent and puerile: the idea of a game, and of loss and gain, does not befit the gravity of the subject [1734, p. 32].

9

THE ART OF THINKING

Chances, odds, 'hazards' (the stock in trade of aleatory probability) are basically quantitative. There is no way to understand odds without understanding numerical ratios. Epistemic probability is not like this. You can compare the degree to which evidence warrants several propositions without recourse to numbers. Indeed Keynes argued masterfully in Chapter 3 of his *Treatise on Probability* that many comparisons of probability are necessarily qualitative and cannot be represented by real numbers. Subsequently B. O. Koopman [1940] elaborated the logic of qualitative probability. A new book by T. Fine [1973] advances this work several stages further. There is nothing logically defective in mere comparisons of probability. But as a matter of historical fact epistemic probability did not emerge as a significant concept for logic until people thought of measuring it. When did this begin?

It is convenient to answer by looking at the word 'probability' itself. We need not do this. We are concerned with the first occasion on which some probabilistic expression with epistemic overtones was systematically used to denote something measurable. We could survey the usage of 'credibility' or whatever, but in fact the word 'probability' itself is the one to watch for. So we ask the exact question: when was this word first used to denote something measurable?

The answer seems to be 1662, in the concluding pages of the Port Royal *Logic*. There could not have been a more auspicious beginning. *La logique, ou l'art de penser* was the most successful logic book of the time and cast the mould for generations of future treatises. Scholars distinguish five editions of the book, beginning with one edition of three printings (in different formats) in 1662, and undergoing minor changes until the fifth and final edition of 1683. All the nations of Europe had their translations and the book

was still used as a text in nineteenth-century Oxford and Edinburgh.
Pascal's associates at Port Royal wrote the book. There are
conflicting reports on authorship, chiefly concerning the proportions contributed by Pierre Nicole and Antoine Arnauld. It is
generally agreed that the latter did more than the former, and in
particular wrote the whole of Book IV that interests us. In his time
Arnauld, the most notable member of a family of influential and
learned men, was distinguished as 'the great Arnauld'. When he was
comparatively young he was asked to comment on Descartes'
Meditations. His chief criticism – the so-called Cartesian circle – is
today still the crux of Cartesian interpretation. When Arnauld was
older Leibniz sent him a sketch of a *Discourse on Metaphysics.*
Arnauld's query about Leibniz's conceptions of substance
and of human freedom produced a correspondence that
remains the central document for understanding Leibniz's system.
Yet undoubtedly Arnauld's chief contribution to philosophy was
something else again: the Port Royal *Grammar.* We have recently
become aware that grammar was one of the chief preoccupations of
the metaphysicians of the time, and no work was more important
than Arnauld's *General and Reasoned Grammar.*

Arnauld's philosophical acumen is attested on all sides but we
may still wonder whether he wrote the final 'probability' chapters of
the *Logic.* There exists a manuscript of the book almost certainly
antedating the published version, and it does not include the four
final chapters. These have not simply been lost, for the table of
contents fails to mention them. Perhaps when the manuscript was
written around 1660 the chapters did not exist or were not deemed
suitable for inclusion. The evidence is inconclusive, for as the
Preface to the *Logic* tells us, several pirated manuscripts were in
circulation; however the *Bibliothèque Nationale* manuscript does
appear to be a legitimate predecessor of the published version.
Scholars have not (to my knowledge) assessed whether the four
'probability' chapters are by the same hand as the rest of Book IV.
To an amateur eye they certainly do not look the same as what went
before.

Whoever wrote the probability chapters started something new.
The first three books of the *Logic* cover Conception, Judgement and
Reasoning; the fourth book is about Order. This is a fairly standard
arrangement. 'Reasoning' deals with the syllogism, 'Order' with
deductive non-syllogistic reasoning such as characterizes most

mathematics. In particular the first ten chapters of Book IV are an elaborate discussion of the notions of 'analysis' and 'synthesis' which, from classical times, had been supposed to constitute the two kinds of geometrical inference. Then there is a lame and conventional chapter on what we can know not through demonstration but through faith. That ends the manuscript version of the *Logic*.

The additional chapters, which I have been calling the 'probability' chapters, begin the study of a novel kind of non-deductive inference. The theory of inference found in the rest of the *Logic* has a strongly Cartesian bent. It is not based on the new Cartesian method which starts with hyperbolic doubt, and which is expounded in the *Discourse on Method* and the *Meditations*. It is rather along the lines of the older *Rules for the Direction of the Mind* which Descartes did not publish, but of which Port Royal had a manuscript copy. The *Rules* have no discussion of non-deductive inference. Until its final chapters *Logic, or the Art of Thinking* has none either.

It is important to distinguish two broad classes of non-deductive reasoning. On the one hand there is inference and decision under uncertainty, and on the other there is 'theorizing'. C. S. Peirce marked such a distinction by calling the former induction and the latter abduction. Theorizing, or abduction, concerns the speculative creation of abstract theory to explain phenomena, together with the testing of such theories by their fit with old facts and their prediction of new discoveries. A coarse but effective sketch of the methodology of theorizing is to be found in Karl Popper's *Logic of Scientific Discovery*. As long ago as 1843 A. A. Cournot urged that there was a special kind of probability appropriate to theories, and from time to time subsequent philosophers have attempted quantitative explications of this idea. None has succeeded. Our confidence in theories does not seem amenable to any probabilistic treatment. Inference and decision under uncertainty, in contrast, are specifically probabilistic. Our theories are for the time being fixed. With data before us we do not know what has happened, or, what will happen, or, what is the case right now, or, what is the most viable generalization from our data. The class of possible hypotheses is determined, and we can apply probability calculations within that space. The final chapters of *Logic, or the Art of Thinking* are the first public delineation of that problem.

Of course there is no sharp distinction between induction and abduction, between inference under uncertainty and theorizing.

The distinction can be sharpened by contrasting the *Logic* with work by Francis Bacon. It is often said that the latter wrote the first modern treatment of induction, but here we must be careful. He certainly never advocated induction by simple enumeration – in which one lists a bunch of *A* with property *B* and concludes that all *A* are *B*. Bacon has no use for such reasoning. As he says in Section 105 of his *Novum Organum* 'The induction that proceeds by simple enumeration is puerile.' Bacon wanted to get beyond the data of sense by constructing abstract models of the world. He calls that induction. He believes sound theories will be suggested to the scientist only if he makes a grand catalogue of phenomena, but Bacon does not aim at inference under uncertainty. He aims at the construction of novel and deep theories that will explain the inchoate data of sense. The word 'induction' is confusing, for Bacon called such theorizing induction. After Hume many people came to reserve the word for something different. Thus Richard Price, introducing the now famous probability theorem of Thomas Bayes [1763], states that Bayes' solution of a 'problem in the doctrine of chances' is

necessary to be solved in order to assure foundation for all our reasoning concerning past facts, and what is likely to be hereafter [. . .] it is necessary to be considered by any one who would give a clear account of the strength of *analogical* or *inductive* reasoning [1763, p. 371f].

The inductive reasoning that Price had in mind was intended as a deliberate and conscious response to Hume. It is inference under uncertainty whose simplest case is induction by simple enumeration, and has very little to do with Bacon's kind of induction. This is not to say that Bacon's use of the word 'induction' was forgotten. On the contrary, it is preserved in William Whewell's mid-nineteenth century *Philosophy* and *History of the Inductive Sciences*. We find it today in L. J. Cohen's *The Implications of Induction*, a work avowedly guided by Bacon and Whewell, which has little use for probabilistic reasoning. At the turn of the last century C. S. Peirce sought to clarify matters by using the word 'abduction' for Bacon's enterprise and 'induction' for Hume's. This has not caught on. However we use the word 'induction', it is plain that Bacon had little interest in Humian induction and no concern with probability. That branch of the philosophy of science now called 'probability and induction' begins with the Port Royal *Logic*.

Book IV has four probability chapters. Chapter 13 (Chapter 12 in the first edition) has a 'rule for the proper use of reason in determining when to accept human authority'. Chapter 14 applies the rule to miracles, Chapter 15 to historical events and Chapter 16 to future contingent events. In this final chapter the author applies numerical measures to probability. One 'contingent event' in question is the winning of a game where each of ten players risks one coin for an even chance of getting ten back. Loss is *neuf fois plus probable* than gain. There are 'nine degrees of probability of losing a coin for only one of gaining nine'. These are the first occasions in print where probability, so called, is measured.

The author not only counted degrees of probability but also knew how to use them. He grants that gaming is trivial but immediately points out that the same kind of reasoning can be applied to many practical considerations. There are for example

many people who are excessively terrified when they hear thunder [. . .] if it is only the danger of death that fills them with their extraordinary fear, it is easy to show that this is unreasonable. It would be an exaggeration to say that one in two million people is killed by a thunderstorm; there is scarcely any kind of violent death less common. Fear of harm ought to be proportional not merely to the gravity of the harm, but also the probability of the event, and since there is scarcely any kind of death more rare than death by thunderstorm, there is hardly any which ought to occasion less fear.

This passage shows that the author is willing to use frequency to measure probabilities of natural occurrences, and it shows he is well aware that a decision problem requires a calculation of expectation involving not only utility but also probability. 'We must reorient', he says, 'many people who conduct their lives as if they should avoid business which may have a dangerous outcome and prefer affairs which may have advantageous results. We ought to fear or hope for an event not solely in proportion to the advantage or disadvantage but also with some consideration of the likelihood of the occurrence.' There is only one case in which the probability of an event is irrelevant to deciding what to do. So long as the probabilities are not zero a strategy with infinite pay-off will always dominate all others. Since 'even the slightest chance of salvation is worth more than all the goods of the world heaped together', salvation has infinite utility. Here follows a brief statement of Pascal's argument from dominating expectation, and so ends the Port Royal *Logic*.

The preceding chapter, on historical rather than future events, also applies quantitative considerations. Take the question of whether a contract witnessed by two notaries has been post-dated. Since it is certain that 999 out of 1000 duly notarized contracts have been properly dated, then 'if we know no other particulars about the contract' we ought to believe the dating is honest. 'It is incomparably more probable that the contract before me is one of the 999 rather than the single one of the 1000 that is post-dated.' However, if we learn that the notaries are unscrupulous the document becomes less credible. If we learn that one of the parties to the contract is thereby reported to have lent £20 000, and yet at the time of dating had only £100, then 'I should believe that there was something false about the contract.'

Note how we get a quantitative probability on the first datum but other data merely weaken this quantity in a qualitative fashion until it simply vanishes. No one has yet discovered how to do this numerically unless it be Thomas Bayes, a century after the publication of the *Logic*. Fifty years later Jacques Bernoulli tried valiantly in that part of his *Ars conjectandi* that was intended as his continuation of the *Ars cogitandi* (the *Art of Thinking*'s Latin title). This problem of mixed evidence, part counting for the hypothesis and part against, is still an open one.

The logic applies the model of the notaries to cases of disputed history and, in a preceding chapter, to miracles. In that discussion there is careful attention to the credibility of witnesses. Some tales of miraculous intervention need not be believed, for their authors are so full of fables that their testimony assures us of nothing. In contrast we can credit St Augustine. Even if we are sceptical about the particulars he relates, we know that in addition to his established probity 'it is impossible that a judicious man would have attempted to lie about such a public matter, since many people would have found out his lie and so would have brought nothing but disgrace on the Christian religion'.

The topic of miracles will recur in our history. Naturally such discussions do not begin with our logic but a new tone is set. Observe the contrast with the Jesuit doctrine of probabilism described in Chapter 3 above. The Fathers of the Church do not always agree: there begins casuistry and probabilism. The Jesuits said that when authorities disagree we are free to choose the most 'probable' authority, i.e. the one that commends itself to our moral and practical sense. Pascal's *Provincial Letters* were written to defend Arnauld against the assaults of the Jesuits. The sixth letter on probabilism defends Arnauld's opinions from probabilist arguments. The rule of Chapter 13 of the *Logic* is also part of this campaign. 'How shall I decide to believe in the occurrence of the one rather than the other of two contrary events if I judge them both possible?' The rule is:

> In order to judge of the truth of some event and to decide whether or not to believe in its occurrence, the event need not be considered in isolation – such as a proposition of geometry would be; rather all the circumstances of the event, both internal and external should be considered.

The goal is a calculus for combining evidence to discover which proposition has an acceptable level of probability. It is opposed to casuist arguments where, after settling that a proposition is possible, and compatible with some source of doctrine, we consider whether belief in the proposition is 'approvable' – probable – in terms of its consequences.

The *Logic*'s rule just quoted is the maxim employed in all subsequent discussion of miracles. The 'internal' circumstances of the event are those that bear on the place of the event in nature – whether it is the sort of thing that tends to happen. The 'external' circumstances are 'those that pertain to the persons by whose testimony we are led to believe in the occurrence of the event'. Hume was able to turn this chapter of the *Logic* on its head. In his essay *On Miracles* he argued that no external circumstances could ever suffice to render probable an event improbable enough to be called a miracle. That thought created another flurry in the concept of probability. The Port Royal *Logic* had already set the conceptual scheme in which the debate was to be conducted eighty years later.

The distinction between 'internal' and 'external' matters far more than miracles. Probability became possible only when signs were turned into internal evidence as described above in Chapter 5. It is some confirmation of this thesis that in the *Logic*, within a few pages of the first measurement of 'probability,' there is a new and explicit statement of the distinction between internal and external evidence. The *Logic* is an anonymous work – we know the hands that wrote it, but not which hand wrote what. That is appropriate. The clear and distinct apprehension of internal evidence is not the work of an individual but of the age. Wherever we look we shall find this new awareness. It may be useful to illustrate by taking as example some individual personage, evolving over a few decades, yet in perfect harmony with that bourbaki of writers, the authors of the Port Royal *Logic* and *Grammar.*

John Wilkins (1614–72), Bishop of Chester, was, with Oldenburg, the first secretary of the Royal Society. What began in, among other places, his own rooms in Wadham College, was, by his talent for organization, transformed from a select but casual club for the best minds of the time into the first and foremost of scientific academies. Four disparate books of his merit attention. Two were written about 1640, two in the late 1660s. The former are worthless; the latter, influential. The difference is not that of an immature young man gaining stature, but of certain ideas visibly maturing in the discourse of the time. One early book, one late, is about language; one early book, one late, consists of probable argument.

Language became exciting in the middle of the seventeenth century. People began to believe that an understanding of language would resolve the chief philosophical problems about the nature of the world. Earlier, the Renaissance aimed at some Urlanguage, writ on the firmament, spoken by Adam and maybe the prophets, and inscribed by signature on every leaf and stone. But as the concept of internal evidence begins to present itself, it furnishes a concept within which to embed the 'signs' furnished by nature, hitherto part of the Urlanguage. All that is left for language is human speech. This is conventional. Hobbes had the full courage of conventionality, but he is unique. Less bold spirits thought there must be a true language, or rather a 'real characteristic' or 'universal characteristic', whose simple elements match the simple elements of nature, and whose rules for compounding will generate each of the possible worlds that nature admits. The most ambitious and most memorable contributor to this programme is Leibniz. The idea and the very term 'real [or universal] characteristic' is not that of Leibniz but, among others, of Wilkins. Louis Couturat's index to *La logique de Leibniz* directs us to ample evidence for this. Leibniz felt the need to assert that he had the germ of the notion of a universal characteristic before Wilkins [*P.S.* III, p. 216], insisting that it is implicit in his 1665 *Art of combinations.* It is now quite usual to say that it all began with the alchemist logician, Raymond Lulle, who died in 1315. The idea was certainly in the air when in 1629 Mersenne described to

Descartes some unknown person's plans for a universal language. Descartes said such a project could not succeed until one had the 'true philosophy'. So in 1629 it had not yet become apparent that this very language might be the way in which to do the true philosophy. The notes that Leibniz scribbled on his copy of Descartes' letter show how radical is the change in the course of a few decades. Leibniz thinks you cannot advance the true philosophy without the sketch of a universal characteristic [Couturat p. 176].

Wilkins is a less Olympian figure whose own mediocre work tells the same story. In 1641 he published *Mercury, or the Secret and Swift messenger, showing how a man may with privacy and speed communicate his thoughts to a friend at a distance.* Mercury, it transpires, is a code together with mechanical suggestions for telegraphic semaphore. Yet the very invention of codes, banal in itself, was pregnant. The preceding year had seen Hobbes saying all language is conventional. Let us then by convention construct a semaphore of 'real signs' that will supersede our worthless hunt for an Urlanguage! In 1661 George Dalgarno published, with ample citation of Wilkins, his *Ars signorum, vulgo character universalis et lingua philosophica.* Its successor, in 1668, was Wilkins' *Essay Towards A Real Character and a Philosophical Language.*

These enterprises are the practical effect of the thesis that all language is conventional. They also represent the antithesis, that language must in some way correspond to the world. Those fond of paradox could become Hobbists, but the serious aimed at a synthesis, namely, the real characteristic, all convention, yet all correspondence. In the Renaissance there were signs, real signs, written by God on nature. People spoke with signs, but so did the world around us. The testimony of man and of nature was one. Then the sign became divided into 'natural' and 'arbitrary'. Hence the desperate plunge into 'real characteristic' which would conjoin, if only in constant conjunction, what the atheist Hobbes had put asunder. But just as there was required a theory of the conventional side of signs, so there was needed a theory of their natural side, which is internal evidence and probability. I have used Wilkins' transition from the pedestrian *Mercury* to the more influential *Real Character* as a symptom of the new awareness of conventional language between 1641 and 1668. Another pair of his works will indicate the rise in probability that went, of necessity, hand in hand with the new theories of language.

The Discovery of a World in the Moone or a discourse tending to prove that 'tis probable there may be another habitable world in that planet, published by Wilkins in 1640, is an essay in the use of probable evidence. Wilkins was a romantic. He had schemes for submarine exploration of the polar icecaps. *Nautilus* had to wait, but probable argument is around the corner and Wilkins seems to know it. His speculations are presented in a sequence of propositions in the style of geometers, but their content is probabilistic. Take no. 13: 'That 'tis probable there may be inhabitants in this other world [the moon] but of what kind they are is uncertain.' According to the preface the arguments for such propositions shall be 'most probable to thy reason'. The reader, not authority, has become the owner of probability. Despite this, much of the argument proceeds by citing authority. It is the testimony of others that establishes probability, and there is more than a taint of probabilism in this work. The authority is sometimes that of Galileo, but the method of argument is casuistry. Indeed proposition 13 is chiefly defended on the ground that some authorities say life on the moon is not impossible, and, if such life be possible, then a beneficent God would surely have planted life where he could.

The ephemeral *Mercury* of 1641 bears a relationship to the important *Real Characteristic* of 1668 similar to that of the mischievous *Moon* of 1640 to a work of Wilkins published only after his death in 1672: *Of the Principles and Duties of Natural Religion.* *Mercury's* code turned into the philosophical language for comprehending all nature. The *Moon's* probable arguments became a general theory of probable argument that, as J. H. Bernard [1896] has shown, is brought to fulfilment in Joseph Butler's 1736 *Analogy of Religion.* Where Butler coined his celebrated aphorism, 'probability is the very guide in life', Wilkins had written,

In all the ordinary affairs of life men are used to guide their actions by this rule, namely to incline to that which is most probable and likely when they cannot attain any clear unquestionable certainty [1675, p. 30].

Wilkins is the first representative of what I shall later call 'Royal Society theology' and whose most familiar doctrine is the argument from design for the existence of God. The argument proposes that the way things are, constitutes evidence for the existence of a supreme being. A universe so well constituted could only be the work of a sublime artisan. Aquinas also had a teleological argument, but his is quite different. It supposes that everything in the world is acting purposefully, and so seeks an agent to endow matter with intentions. In the Thomistic teleological argument, the world is an agent, which, since it is not in itself active, demands a Supreme Agent. In the argument from design, the world is passive. We hunt not for an agent but an artisan. The world is not 'external evidence' written by God, but 'internal evidence' which can be explained only by the existence of a God. In confirmation of this thesis we find that Wilkins, first propounder of an argument from design, begins by presenting the new categories of evidence.

Evidence is of three kinds. Two are familiar to the author of *Moone*: demonstration and testimony. But then there is another category, which Wilkins can only call 'mixed', a half-way house between testimony and demonstration, for 'besides these there is a mixed kind of evidence relating both to the senses and the understanding, depending upon our own observation and repeated trials of the issues and events of actions or things, called experience'.

Wilkins may be our best witness to the fact that what Port Royal called 'internal' evidence is new. In twentieth century epistemology there is only one fundamental kind of evidence, namely internal evidence. In the mid seventeenth century this kind of evidence could make its way only as a sort of wedge between demonstration and testimony. Yet the decisive feature of this wedge is that, by being conceived of as *in the middle*, it united what had hitherto been utterly disparate. I have argued that the objects of knowledge were typically different in kind from the objects of opinion. Things known were not especially well supported things opined, for things known had evidence of a quite different sort from that furnished for opinion. But Wilkins enunciates the view that, as he matured, was becoming commonplace. 'Evidence less plain and less clear is probable.' Probable evidence enters the same league as demonstration.

It is of particular importance that although Wilkins in his youth was dreaming of expeditions to the moon and the antipodes, he was, in his later years, bringing God down to earth. In the Thomistic theology, God, insofar as He could be conceived, was an object of knowledge, not opinion. Pascal had dared to make God an object of decision, not belief. Wilkins, ushering in a more complacent age, dared to make God an object of probable opinion. Beliefs about God were now in the same category as the belief that the sun will rise tomorrow, or, to take another of Wilkins' examples, the belief that my house will not fall in tonight. Pascal's wager, presumably not unknown to Wilkins, is demeaned. We are to act according to some balance between utility and probability; the greater the utility, the less the probability required; the argument from design furnishes enough probability of God's existence for us to be Deists.

10

PROBABILITY AND THE LAW

No particular event, no single student, is responsible for the emergence of probability. In order to represent epistemic probability on a numerical scale the Port Royal *Logic* used gaming as its model. It may be tempting to infer that probability could be measured only by using concepts devised for games of chance. That would be a mistake. A gifted mind ignorant of the doctrine of chances but able to apprehend the fact that evidence and causation are in different categories could perfectly well start measuring epistemic probability. The proof of this is that Leibniz did.

In 1665 when he was 19 he published a paper using numbers to represent what he called degrees of probability. At that time he was isolated from the mainstream of European thought and he did not know much mathematics. His best teacher of that subject had been Erhard Weigel at Jena in 1663. Weigel certainly knew no theory of chances; indeed, he had not even mastered Cartesian geometry. Yet Weigel was a significant teacher. He held the commonplace idea that all knowledge should be axiomatized on a Euclidean plan, but he coupled this with a peculiar faith in arithmetic. He liked to toy with eccentric equations and non-standard arithmetics – he was especially proud of his system to the base 4. Leibniz's combinatorial imaginativeness was undoubtedly supplemented by this kind of tutelage.

In addition to being a mathematician Weigel was notably learned in jurisprudence and had various projects for making a deductive science of law. Hence he was an appropriate teacher for Leibniz to seek out. Leibniz's own education had been chiefly legal. Law was in the family. His father was professor of moral science with some reputation in jurisprudence. His mother's father was a professor of legal theory. Philosophical stories about the young Leibniz relate how he skimmed Aristotle and the scholastics in his father's library.

He probably spent more time on the ample store of legal books. His doctoral thesis for Nuremburg dealt with perplexing cases in law. It was rejected, it is said, because the author at 20 was too young. Nearby Altdorf snapped him up and offered him a law professorship. He declined in order to enter the world of affairs but he was ever a lawyer, compiling both general theses on jurisprudence and particular briefs for his various employers. This hack work with its relative lack of success did not diminish his respect for the subject. Long after he had mastered mathematics and even co-invented the differential calculus Leibniz advanced 'a paradox which, though amusing, gets at the truth: there are no authors whose style is more akin to the geometers than the old Roman jurists in the Pandects' [*P.S.* vii, p. 167]. The Pandects – the great digests of Roman law compiled at the decree of Justinian – formed, with their massive commentaries, a chief element of Leibniz's legal education.

Probability is not foreign to the law. Evidence, the stock in trade of the epistemologists, is primarily a legal notion, although it took its present place in modern European systems surprisingly late. The concept of epistemic probability requires us to recognize differences between what causes things to happen and what tells us that they happen. Only one of the professions stuck fairly fast to this distinction: civil law. The advocate must distinguish testimony from circumstance. So Roman law had some paraphernalia of scales of evidence. When Leibniz finally got clear about the role of a theory of probability, he called it 'natural jurisprudence' [*P.S.* iii, p. 194]. Or, to reverse the direction of the metaphor, 'The whole of judicial procedure is nothing but a kind of logic applied to questions of law' [*N.E.* iv. xvi. 9]. Mathematics is the model for reasoning about necessary truths, but jurisprudence must be our model when we deliberate about contingencies.

Once again it is convenient to contrast Locke and Leibniz. The former in his *Essay* fashioned a discussion of degrees of assent on the corresponding part of the Port Royal *Logic*. He is more cautious than his predecessors. Probability, for Locke, is 'likeness to be true, the very notation of the word signifying such a proposition for which there be arguments or proofs to make it pass or be received for true' [iv. xv. 3]. When all experience and testimony coincides, 'probability', he says, 'leaves as little liberty to believe or disbelieve as a demonstration'. Unfortunately testimony often conflicts and there is what Port Royal called disagreement between internal and external evidence – that is, the stories of witnesses may go against the common run of things. Such cases

are liable to so great a variety of contrary observations, circumstances, reports, different qualifications, tempers, designs, oversights etc., of the reporters that *it is impossible to reduce to precise rules the various degrees wherein men give their assent* [italics added].

Leibniz faithfully reports Locke's defeatist doctrine and retorts that lawyers already have a whole family of just such rules for degrees of assent, or at least of proof. The examples Leibniz offers in the *New Essays* are not very impressive, but he had done better long before, in his Baccalaureate essay of 1665, *De conditionibus*. This was revised for inclusion in a *Specimena juris* in 1672, and much later he thought it should be 'retouched' and published again. I cannot tell how good a piece of work it is. It seems to be a curious mixture of logical insights combined with a jejune and oversimplified view of legal process.

De conditionibus is a study of conditional rights. In law my right to something may be absolute (*purum*) or absolutely void (*nullum*) or it may be conditional. The settling of estates with conditional rights and liens attached to them is a standard legal problem. Leibniz attacks it with the abstractness and axiomatic form he learned from Weigel, I shall try to explain it using more concrete illustrations. Let *r* be the proposition that a person has a certain right, say, to some property. Let *q* be the condition that his uncle has willed this land to him. Then the condition may be expressed by the sentence 'if *q*, then *r*'. Leibniz's investigation of conditional rights is tantamount to a study of those propositions that logicians now call conditionals or hypotheticals, and which can be written in the form 'if . . . , then . . .' His study of conditionals is of independent interest, but here we notice only how it leads him to a theory of what we might now call 'partial implication'.

A may make a will that leaves his property for the use of *B*, on whose death the property is entailed to *C* unless *B* has issue. Let *r* be the statement that *C* has the right to the property, and let *q* state that *A* made such a will. Then it is not true that if *q* then *r*. Let p_1 be the proposition that *B* has living descendants, and p_2 the opposite. Then *q* essentially subdivides into $(p_1\text{-or-}p_2)$. If p_2, then *r*, but if p_1, then not-*r*. This is the simplest sort of case which might interest a lawyer who has to realize some property.

In general, consider propositions of the form 'if q then r' where q is a disjunction of mutually exclusive alternatives. Now consider a set of conditions each of which is sufficient for r. Three cases may arise. Every disjunct of q may preclude each of these conditions. In that case Leibniz calls the condition for r impossible, and we have *jus nullum*, no right at all. If every element of q entails some condition sufficient for r, then the condition is called necessary, and we have *jus purum*. However if some disjuncts entail a condition for r, while the rest entail a condition for not-r, then we have only a conditional right, and the condition is called uncertain (*incerta*, in the 1665 version) or contingent (*contingens*, in the 1672 version). When, in q, the conditions for r are uncertain, part of q favours r and the rest favours the opposite. So we have a sort of partial implication: part of q implies r. When the implication is complete (that is, when the condition is necessary) Leibniz denotes it by the figure 1; when the condition is impossible, he uses the cypher 0; when the condition is uncertain, the implication must be denoted by a fraction. Moreover these fractions denote what are variously called 'the degrees of proof' for the right, or its 'degree of probability'.

Leibniz made no concerted attempt to evaluate these fractional degrees of probability. He could have been interested in that problem, if he had been a lawyer with instructions to divide an estate before it is known which conditions are fulfilled. But in fact he was interested in theorems about *jus purum*, that is, he wanted to know what kinds of combinations of conditions, none complete in themselves, would justify a conclusion of unconditional right. He even put this theory to work when in 1669 he had to prepare a brief on that bizarre intertwining of conditional rights, the disputed throne of Poland.

Generalizations of his idea came very quickly. He set to work on *Ad stateram juris, de gradibus probationum et probabiliatum*. The degrees of proof and probability of which the title speaks are to be applied quite generally, but must always take jurisprudence as their model. The essay begins with a ringing denunciation of that rival theory of probability, the probabilism of the Jesuits. Evidently Leibniz had been reading Pascal's *Provincial Letters* and this exposition echoes them frequently. Often in the course of his life he thought it necessary to distinguish casuistical from real probability: 'I do not speak here of the probability of the Casuists, which is founded on the number and reputation of the Doctors, but that which is derived from the nature of things, in proportion to what we know about them' [*P.S.* VII, p. 167].

At the time that Leibniz was beginning to think about his 'new kind of logic' based on degrees of probability, he was working on the better known *Art of Combinations*. This has now been canonized as the first entry in Alonzo Church's definitive *Bibliography of Symbolic Logic*. Leibniz did not at first see the relevance of his two enterprises: he was unaware that the probability theory would for long be a combinatorial discipline. This contrasts with Pascal's 1665 *Arithmetic Triangle* whose theory of combinations is chiefly motivated by problems in chances. There has been some speculation as to whether Leibniz knew Pascal's book. It seems very unlikely for Leibniz does not draw the connection between probability and combinations until after his long stay in Paris, 1672–6, where he mastered the work of 'Pascal, Huygens and others'. Much later he recognized how closely probability and combinations are connected; when he was drafting a plan for a new edition of *De arte combinatoria* he wanted the appendix to include the best statistical data then available, namely John Hudde's tables of mortality [*Cout.* p. 561].

When Leibniz first wrote on combinations he had a quite different aim. Suppose you would have a 'flash alphabet of human thoughts', that is, a list of all the words denoting simple ideas. Then any complex idea could be formed from these by combination. A marvellous 'Art of Invention' would result. All possible ideas and all possible propositions would be generated mechanically, so that we would be able to survey not only what we know but also what we do not know, and hence conduct deeper investigations. This project obsessed Leibniz throughout his life.

Leibniz did not contribute to probability mathematics but his conceptualization of it did have lasting impact. Most of his contemporaries started with random phenomena – gaming or mortality – and made some leap of imagination, speculating that the doctrine of chances could be transferred to other cases of inference under uncertainty. Leibniz took numerical probability as a primarily epistemic notion. Degrees of probability are degrees of certainty. So he takes the doctrine of chances not to be about physical characteristics of gambling set-ups but about our knowledge of those set-ups. When he went to Paris he found a mathematics tailor-made for his nascent epistemic logic. In the next chapter, when we look at Huygens, we shall find his book is entirely about games of chance and has few epistemic overtones. The word 'probability' does not occur. Leibniz, however, could call it 'an elegant example of reasoning about degrees of probability' [*Dutens* VI. i, p. 318].

From the beginning Leibniz thought of probability theory as the logic for contingent events. We describe his inductive logic later. Here let us note just two of the consequences of regarding inductive logic as 'natural jurisprudence'. The concept of conditional probability – the probability of *A* given *B*, now written *Prob*(*A*/*B*) – was slow to evolve and lacked a perspicuous notation until quite recently. Lack of this concept can make reasoning difficult and is conducive to error. Some philosophers and statisticians contend that in fact all probability is conditional. Such an opinion does not naturally occur when you are thinking chiefly about games of chance. Only if one has the epistemic point of view is it attractive. In legal process all inference is relative to or conditional on the evidence made available to the court, so Leibniz, with law as his model, took for granted that probability is 'in proportion to what we know'. Or, as he very often writes, all probability conclusions are *ex datis*, relative to and derived from the given facts [e.g. *P.S.* VII, p. 201]. The whole point of probability is that we may not be able to establish a proposition with certainty; we can at best measure the extent to which the data warrant our inferences. Leibniz owed his awareness of the conditional character of epistemic probability to his legal model.

For a second and more subtle influence of the law, let us reconsider the problem of mixed evidence. Suppose that some evidence counts for *r* and some counts against it. Nowadays most of us accept that the relation between some propositions and others may be fundamentally probabilistic. We acknowledge this because we have to accept as an irreducible fact, that a quantum system in state Ψ confers a 60% probability on the proposition that a particle is in region *r*. According to the formalism of quantum mechanics Ψ cannot be decomposed into Ψ_1 or Ψ_2 with Ψ_1 entailing that the particle is in *r* and Ψ_2 entailing the opposite. Of course very similar facts have been familiar in everyday life, although only microphysics makes some people notice them. The fact, ϕ, that the rear tires are completely bald, confers 60% probability on having a flat before reaching Massachusetts. It is a mere myth that ϕ should break down into a set of cases ϕ_1, which entail getting a flat tire, and another set ϕ_2, entailing the opposite. We certainly know of no such decomposition and have no good reason to think there is one. Yet the myth that every problem in probability can be reduced to a set of favourable and unfavourable cases persisted for centuries. One reason for this is obvious: the analysis fits many games of chance. But there is another source of the idea. The analysis also fits the theory of conditional rights. No rights are irreducibly probabilistic (unless a wit instructs that his estate is to be apportioned by lottery). If I have only a conditional right to r, then there exists some disjunction of conditions any of which, when fulfilled, suffice for *jus purum*. There is another set with the opposite meaning. Thus the principle of analysis into favourable and unfavourable cases is sound in the law of right. Leibniz, taking law as his model for probability, was over-ready to accept the analysis into cases. (It was also overly easy, for as we shall see Huygens took *casus* to render his Dutch *kans* or chance – and this quite rightly, after the Roman form *casu* for events that happen involuntarily. Thus an analysis into equal chances was, by an unavoidable punning, an analysis into equal cases). Leibniz certainly did nothing to dissuade Jacques Bernoulli from following the analysis into cases. Indeed when Bernoulli considers the problem of annuities he follows Leibnizian legal terminology. When all alternatives point in one direction, we have what Bernoulli called the *pure* situation; when there is conflict, the situation is 'mixed'.

11

EXPECTATION

It may seem as if mathematical expectation should have been easier to grasp than probability. From an aleatory point of view the expectation is just the average pay-off in a long run of similar gambles. We can actually 'see' the profits or losses of a persistent gamble. We naturally translate the total into average gain and thereby 'observe' the expectation even more readily than the probability. However the very concept of averaging is a new one and before 1650 most people could not observe an average because they did not take averages. Certainly a gambler could notice that one strategy is in Galileo's words 'more advantageous' than another but there is a gap between this and the quantitative knowledge of mathematical expectation.

Cardano's notion of 'equality' and 'the circuit' in games of dice is some anticipation of mathematical expectation but it is difficult to follow in detail. Not until the correspondence between Fermat and Pascal do we find expectation well understood. This concept is at the very heart of Pascal's wager. Recall, however, that the Port Royal *Logic* thinks it important to 'reorient' people so that they base decisions on both utility and probability. This suggests that a comprehension of expectation was not something one could take for granted even in 1662. Yet shortly before there had been a really thorough statement of concepts akin to expectation. They are well worth scrutiny. I refer to the first printed textbook of probability, Christian Huygens 1657 *Calculating in Games of Chance*.

Holland was briefly a chief centre for the new Cartesian mathematics. Descartes himself had settled there and his geometry was mastered not only by scholars but also by men of affairs such as John Hudde or John de Witt, respectively mayor of Amsterdam and Grand Pensionary of the Netherlands. All the same, Paris remained the intellectual capital of Europe where one went to finish one's education. On Huygens' second expedition to that city he heard about the Pascal–Fermat correspondence and made the acquaintance of Roberval, Carcavi and Mylon, all of whom knew the protagonists. Subsequently Huygens became a close associate of Roannez, and even met Méré, whom in his journal he calls the inventor of the division problem. In 1655 Fermat lived out in the country and Huygens believed that Pascal, now in Port Royal, no longer contemplated mathematics. So the young Huygens met neither of these heroes, nor, it seems, did he see any written solution of Méré's problems. He must have heard the gist of the solutions but he went home and worked them out for himself.

F. van Schooten, a Cartesian mathematician and publisher, begged Huygens to write up these results for a forthcoming series of mathematical tracts. Huygens obliged. In 1656 he sent a manuscript to Paris in the hope that Fermat or even Pascal might be got to look at it and approve the solutions. It took four months for a reply but the confirmation of Huygens' work was perfect. Moreover Pascal sent back another problem in chances and Fermat sent two. These together with two further problems devised by Huygens were put at the end of the textbook and for sixty years formed the standard tests against which one displayed one's skill in the doctrine of chances. Abraham de Moivre, Jacques Bernoulli, Nicholas Struyck, Pierre Rémond de Montmort and John Arbuthnot all published solutions to some of these; Hudde, Spinoza and Leibniz are among others who tackled them.

Huygens' monograph first came out in Latin in 1657 under the title *De ratiociniis in aleae ludo*. The Dutch edition did not appear until 1660 as *Van Rekiningh in Spelen van Geluck*. The vernacular version was written first. There was some mild bickering between Huygens and Schooten as to who would do the translating. In the end Huygens sketched out the style of translation and Schooten did the actual work, by no means to Huygens' satisfaction. [Consult the Bibliography for references to items in Huygens' *Oeuvres* pertinent to this and subsequent topics.]

Most seventeenth century writers thought that Latin was the most suitable language for expounding mathematics. Time and again we find a correspondent starting a letter in French or whatever and being forced to break into Latin. The vernaculars were not deemed rich enough in the burgeoning vocabulary of pure mathematics. But the doctrine of chances is applied mathematics arising from vulgar practices and has plenty of terminology. Huygens could express himself in Dutch but had trouble finding suitable Latin terms. The editors of his book have found a page in which he lists all the Latin candidates for various Dutch concepts – *alea, sors, fortuna, casus, lusiones, etc.* These became the standard terms for a century and a half of subsequent publication.

Both of Méré's problems require some sense of the idea of expectation. It was Huygens' practice throughout his work in applied mathematics to follow the precept of Archimedes, and begin any treatise with a set of specific rules or axioms peculiar to the science he was developing. Thus his textbook on chances has the same goal of rigour as has a modern treatise. We look for, and find, a very clear account of ideas of expectation.

Huygens needs to know the value of any particular gamble. That is, if we are invited to gamble with a given schedule of prizes depending on various outcomes, we demand the fair price for taking the gamble. Our standard answer is that a gamble is worth the mathematical expectation of that gamble. Centuries of custom make this answer seem self-evident to us, but it was not the established answer when Huygens wrote. Hence he had to justify mathematical expectation.

Even today justification is called for. Nowadays people tell you that the expectation is the fair price, because if you repeatedly gambled on the same terms, the expectation is your average pay-off. If you paid more than the expectation you would tend to lose, and if you paid less you would tend to make a profit. This is undoubtedly a sound rationale for buying several hundred gross of tickets in successive lotteries, but what if you are going to buy only one ticket in one gamble? Why should long run average profit be the measure of fairness?

Churchill Eisenhart has a nice illustration of this predicament. A Pacific naval base has a machine selling Coca-Cola at 5¢ a bottle. The price of Coke goes up to 6¢. The machine takes only nickels. If at random one in every six bottles in the machine is empty, then the machine is undoubtedly fair. Such a mechanical vendor may be fine for the regular patrons but a casual visitor who has to pay a dime before he gets a drink is not likely to think the game fair even if he is warned beforehand.

It is important to distinguish two distinct questions. On the one hand we may query whether the mathematical expectation is a fair price for a gamble when we will gamble only once. This is the question of pricing. Now even if we answer this question affirmatively, there is a further question of justifying the answer. Examples like Eisenhart's do not really question that the expectation is the fair price of a gamble. They remind us that for unique gambles the long-run justification cannot be right. There do exist amusing cases, of which the St Petersburg problem is the first on record, that call in question the practice of pricing by expectation. [See Todhunter 1865, p. 220.] However, Huygens had no doubt that expectation is the fair price. This left him with the problem of justification. His solution to this has a singularly modern flavour because it has been revived, in a different format, in the personalist theory of F. P. Ramsey and L. J. Savage.

In discussing mathematical expectation we do so with hindsight. There was no ready-made concept conveniently labelled 'expectation' whose use Huygens wanted to justify. He does use the word *expectatio* in his first tentative translation of his Dutch manuscript, and to him we owe the word, although for long *spes*, or hope, ran it a close rival. But Huygens was not trying to justify expectation; he was trying to justify a method for pricing gambles which happens to be the same as what we call mathematical expectation.

Huygens thinks there is one basic situation in which we know the fair price for a gamble. He, Huygens, supposes the lots are perfectly symmetric so each can be drawn 'as easily as' any other. We shall not now concern ourselves with how Huygens' 'ease of drawing' is to be explicated. It is of no moment whether the symmetry is in the physical make-up of the lottery, or in the equal relative frequencies with which the tickets are drawn, or in the sheer subjective indifference of the bettor to which tickets he happens to hold. With our modern sophistication we can divine different bases for the symmetry of the fair lottery – frequency, propensity, personal opinion and so forth – but Huygens cheerfully and perhaps rightly takes symmetry as the primitive and undefined notion.

In a fair lottery it is clear that every bettor must pay the same price for any ticket. Moreover, if the prize is z then each of the n tickets should cost z/n. If the tickets cost more the owner of the lottery would profit without risk. If the tickets cost less, the bettors could form a syndicate which would profit without risk. So if each player pays x then the prize must be nx.

Huygens tacitly takes for granted several principles of utility theory. Fair prices are additive. Lottery tickets are not cheaper by the dozen. Lotteries may also be compounded. Let a be the fair price for ticket 1 in lottery X. Let b be the fair price for ticket 2 in lottery Y whose prize is the sum a. Then b is also the fair price for a lottery like Y, but whose prize, instead of A, is ticket 1 for lottery X. Finally, Huygens thinks that consolation prizes do not change fair prices. Thus once we have established the fair price for a lottery ticket, we do not change the price by adding the rider that the winner shall pay equal consolation prizes to the losers (so long as this total side pay-off is less than the original prize, so that the winner is not out of pocket).

Using these assumptions Huygens can now argue for the fair price of any bet. He invents the device of *equivalent gambles*. To settle the price of a ticket T we must find a fair lottery such that a bettor is indifferent between having T or having a ticket in a certain fair lottery. Since he is indifferent the prices must be the same. Knowing the price of the fair lottery ticket we deduce the price of T. There are two ways in which a lottery can fail to be fair. The prizes may be unequal, and the tickets may not be drawn equally easily. Huygens examines the former first.

Suppose you are offered the following contract: a symmetric lottery has two tickets only and you win a or b according to which ticket is drawn; $b > a$. How much should you pay for this contract? First consider a fair lottery with two tickets and prize $a + b$. The fair price for tickets is $\frac{1}{2}(a + b)$. Moreover, this remains the fair price if the winner is bound to pay the loser a consolation prize of a. Hence this is also the fair price for the contract, for in either case one has an equal chance of winning a or b. This is the content of Huygens' Proposition I.

His second proposition generalizes the argument to any number of equal chances worth a_1, a_2, a_3, ... and so on. Finally his Proposition III turns to unequal chances. These are represented by holding more than one ticket in a fair lottery. Thus suppose there are p chances of winning a and q of winning b. By a more elaborate form of reasoning along the same lines as that for Proposition I, Huygens deduces that the value of such a gamble is $(pa + qb)/(p + q)$.

Such reasoning might be formalized in several different ways. Olav Reiersøl [1968] asserts that Proposition III requires the assumption that 'it is possible to find any number of people willing to take part in an equitable game, and it is not possible to find anybody who is willing to take part in a game which is less favourable than an equitable game'. But since this assumption is false and since it does not occur in Huygens it is perhaps to be avoided, although its intention is clear. It is more instructive to compare Huygens' method of argument to F. P. Ramsey's derivation of probability theory. Ramsey starts with an initial assumption that there is an 'ethically neutral proposition'. Similarly L. J. Savage assumes that there is at least one conceivable set of alternatives between which we would be indifferent. At worst, we can at least imagine a fair coin. Ramsey and Savage were developing a theory of both utility and probability whereas Huygens took utility theory for granted, but otherwise the old method and the new one are similar. Both rely on the device of equivalent gambles.

Huygens was able to derive his results only for rational probabilities. Thanks to work beginning with B. O. Koopman [1940] we know that so long as there is no upper bound in the number of tickets in a fair lottery, we can use Huygens' reasoning to prove a representation theorem by which we get probabilities on the real line. This is of some significance. Richard von Mises once vigorously opposed the 'classical' theories of probability on the ground that they were restricted to the domain of rational numbers. In the case of the classic Huygens we now know that this is not necessarily so.

The fair prices worked out by Huygens are just what we would call the expectations of the corresponding gambles. His approach made expectation a more basic concept than probability, and this remained so for about a century. There was nothing wrong with this practice: it has been elegantly revived by P. Whittle [1970]. Huygens does not use any long-run justification for his 'fair prices'. Perhaps this is partly because averages were not well established as a natural way of representing data. Also Huygens is to some extent neutral between aleatory and epistemic approaches to probability, although he leans towards the former. For instance he takes an example in which I hold three equal coins in one hand and seven in the other, and argues that you should pay five for the privilege of choosing a hand. A modern writer with a frequentist outlook would insist that the coins got into my hands at random. A personalist would say Huygens was merely expressing subjective indifference. But Huygens is simply untroubled by such modern sophistication and thinks his example is clear enough.

Huygens had to make up his terminology as he went along. He chose the word *expectatio* when speaking of the value of a gamble.

In the Dutch original he wrote *kans*. We can infer from other writings of the time that this was not as clear as he would have liked. Spinoza deals with the same problem, the justification of expectation, in a letter of 1666. He begins rather like the Port Royal *Logic*, saying that *kans* is proportional to the *lot* and to the *geld* which is being wagered. Here *kans* appears to be a derived concept, proportional to *lot* (probability?) and *geld* (money). But in the remainder of the letter *lot* drops out and *kans* becomes the dominant concept, with equal probability (for equal pay-off) being called equal *kans* instead of equal *lot*. It rather looks as if Spinoza had tailored his argument to Huygens' book, which is not mentioned, but which is followed in some detail.

The letter shows that Huygens was not alone in seeking a rigorous justification for his system of pricing gambles. Spinoza was writing to one J. van der Meer, who apparently had asked him this very question. No copy of van der Meer's letter is known, nor does anyone hazard a conjecture as to who he was. We possess only a Dutch draft of the letter (the Latin one in the *Opera* is a posthumous translation). It is a wonder that those who think that Chevalier de Méré had a significant role in probability theory have not made the multilingual pun and postulated that Spinoza's letter is a draft for a letter in French to Méré.

The problems at the end of Huygens' book were examined by several generations of probabilists. The questions set by Fermat and Pascal are unequivocal, but both of Huygens' questions are ambiguous. This serves to remind us of the sheer difficulty that even a Huygens had in stating things clearly. For example,

Three players, *A*, *B* and *C* take 12 chips of which 4 are white and 8 black. The winner is whoever first draws a white chip. Given that *A* draws first, then *B*, then *C*, then *A*, and so on, what is the ratio of their chances?

Jacques Bernoulli pointed out that there are at least three different interpretations. First, each time a black chip is drawn it may be put back in the hat. Second, we can have drawing without replacement from one hat. Third, we can suppose that each of the three players begins with his own hat of twelve chips and draws without replacement. It turns out that Huygens had the first interpretation in mind. However, there is a very extensive correspondence with Hudde over this problem. Hudde was the mayor of Amsterdam and an able enough mathematician. In Chapter 13 we shall see that his contribution to annuity theory is of some importance. But he and

Huygens cannot agree on how to solve the problem I have quoted because Hudde puts the second construction on Huygens' words. Neither party can understand what the other is doing.

They also fall out on another question. *A* and *B* are tossing a coin on the following terms. Each time a player gets tails he puts a unit coin in the pot; as soon as a player gets heads he wins the pot. *A* goes first. What is his advantage? Huygens understands that the game will not be over until some money has been put in the pot, i.e. not until after someone has thrown tails. Hudde thinks the game is over after the first heads. They have a row over their solutions to this problem for at first each is unable to see that the other has a correct solution to *a* problem. The possibility of such long drawn out misunderstanding shows how ill-settled is the very language of the probability calculus.

One other example about expectation is instructive: it concerns the phrase 'life expectation'. In 1662 John Graunt used the London Bills of mortality to draw inferences about the death rate in London. Huygens was sent a copy of this book in the same year but at that time he merely expressed admiration for it. Subsequently, in 1671, Huygens was called in to confirm a quite different set of inferences drawn by Hudde and de Witt. In between those times his brother Ludwig, who had been reading Graunt's book, wrote in 1669 asking Christian what is the life expectancy of a new born infant, according to Graunt's tables. Since he does not have any phrase such as life expectancy, he is forced to say, *la question est jusqu'à quel âge doit vivre naturellement un enfant aussitot qu'il est conçue.* That is, to what age ought a newly conceived child live in the natural course of things? 'If you have any difficulties with this problem, I'll let you know my own method, which is guaranteed', says the confident Ludwig.

Christian twitted his brother on directing the question at newly conceived infants rather than new-born ones but Ludwig is half right, for Graunt's tables include the still-born and the aborted among mortalities. Even putting that trifle aside Ludwig's question is ambiguous. Not perceiving the ambiguity, Ludwig worked out what has been called the expectation of life. He first computes, from Graunt's meagre and indeed fictitious data, the relative number of chances of dying at any age. This number is assumed equal to the proportion of Graunt's population who die at the given age. Then, as Christian sums up his brother's notion:

By my rules for games of chance, it is necessary to multiply each number of chances by the number of years corresponding and then divide the sum of the products, which is here 1822, by the sum of the chances, which is here 100. [This is just an application of Huygens' Proposition III, mentioned above.] The quotient, 18 years and about 2½ months, will be the value of the expectation of a newly conceived child.

This number of about 18.2 is not necessarily the answer to Ludwig's question. Although the expectation of life for a newly conceived child is worth 18 years 2½ months, this does not mean that he can be expected to live so long, for indeed it is much more to be expected that the child will die long before the age of 18. For example, although a little over 18 years is the expectation of life, it is certainly not an even bet whether or not a child will, in those shortlived times, live to be 18. The point is elegantly illustrated by Christian Huygens:

Imagine that people were even feebler in their infancy than they are now, and that 90 in a 100 die before they are 6 but that those who exceed this age are veritable Nestors and Methusalehs, and that they live on the average to be 152 years and 2 months.

In this case the expectation of life is once again about 18 years. But anyone who bet that a newborn child would not live past the age of 6 would have an enormous advantage over someone who bet the other way.

The difficulty arose over Ludwig's obscure wording, *jusqu'à quel âge on doit vivre naturellement.* This is because of the ambiguity two distinct concepts explicate these words, now called, say, median age and expected age. Nowadays the great decline in child mortality makes the median and expected age fairly close. But in those days the expected age, on Graunt's data, is 18.2 while the median age is around 11. It appears from the correspondence that although Ludwig worked out the answer 18, the number he actually sought was 11. He had not fixed the concepts in his mind as clearly as his brother.

Both median and expected age convey information. What information is wanted depends on the problem at hand. It is a defect in Ludwig's question that it comes as a riddle and not as a problem. As soon as one has a use for mortality statistics it becomes more clear what information is needed. The first use of such statistics is in life annuities. A life annuity is a contract in which the buyer pays a set sum in exchange for an annuity. Dutch towns regularly sold annuities to raise capital, so these contracts were familiar to Huygens.

Two parameters determine the fair rate for simple life annuities: the prevailing rate of interest and the mortality rate. Joint annuities are mathematically more interesting: if the prevailing rate of interest is 6% and a man of 60 and his wife of 55 jointly pay £1000, what should be their annual return until death of the last surviving partner? Stimulated by his brother's questions, and by investigations then under way from Hudde and Witt, Huygens proceeds to attack such questions with gusto. Here we have the first serious application of the doctrine of chances. Naturally it is effective only when there is some statistical data to which to apply it. Now we must tell that story, going back once again to 1662, when Graunt began to use the London Bills of Mortality.

12

POLITICAL ARITHMETIC

Statistics began as the systematic study of quantitative facts about the state. From 1603 the City of London kept a weekly tally of christenings and burials. Desultory records had existed earlier but a desire to know about the current state of the plague made it necessary to set out the figures in a more regular way. Most of the people 'who constantly took in the weekly bills of mortality, made little other use of them, than to look at the foot, how the burials increased, or decreased; and among the casualties, what had happened rare, and extraordinary, in the week current'. Or so John Graunt [1662] tells us in the preface to his *Natural and Political Observations* upon the selfsame bills. He and William Petty – whose various essays on 'Political Arithmetic' make him the founder of economics – seem to have been the first people to make good use of these population statistics.

Why did no one do so earlier? It is plausible to suppose that inference from statistics evolved slowly because there were few data, but this is not the whole story. It is true that once Graunt had made plain the value of statistics, the capitals of Europe copied London and so data became more ample. For example, Paris started its tabulations in 1667, the year after Petty reviewed Graunt's book in the *Journal des Sçavans* [Petty, 1666]. But plenty of data were already in existence. Annuities had been an established method of national or local fund-raising for a very long time. The records of pay-offs from annuity funds provided ample information about the population. There was a good motive to examine this data, namely to determine whether annuities were profitable to own. But no serious analysis of such material was provided before John de Witt made his presentation to the Estates General of Holland and West Friesland in 1671.

Annuity data were most readily available in the Netherlands, but other records had accumulated elsewhere. For example, only in the past few decades have French and then English demographers begun to decode the registers of parish churches. From these ancient volumes we are beginning to know a great deal about population trends in the sixteenth century. They are not as good as a planned census, but they contain much information. We can now use scores of amateur historians to plough through ancient registers; computers vastly speed up the data analysis; but there is nothing new in principle. Individual registers could have been analysed in 1600, but Petty may have been the first person who did this and told us about it, presenting his results to the Royal Society in 1674. The registers had lain unused, silently amassing information, for a century beforehand. 'All that was needed', wrote one of the first systematic demographers, J. P. Süssmilch, 'was a Columbus who should go further than others in his survey of old and well known reports. That Columbus was Graunt [1741, p. 18].'

It is true that demographic knowledge was of less value to a feudal society than an industrial one. When land and its tillage are the basis of taxation, one need not care exactly how many people there are. As English towns increased in size tax was levied on the most manifest signs of habitation, the number of hearths, and there was still not much need to know the population. But we must also suppose that non-economic factors in the conceptual scheme of earlier times precluded the use of statistical data. Graunt and Petty were ignorant of Pascal and Huygens, but Paris and London, in their very different ways, were simultaneously starting the discipline we now call probability and statistics. Whether motivated by God, or by gaming, or by commerce, or by the law, the same kind of idea emerged simultaneously in many minds.

The bills of mortality used by Graunt were commenced during one of the worst plagues, that of 1603. We have already quoted Fracastoro on the signs of contagion which are 'signs of probability', and which include signs from the planets, the air, and the earth with its insects. He is echoed in one of London's more popular 'defensatives' against the plague, written by Simon Kellwaye in 1593:

When in summer we see great stores of toads creeping on the earth having long tails and an ashen colour on their backs and their bellies spotted and of divers colours, and when we see great stores of gnats to swim on the waters or flying in great companies together [. . .] it showeth the air to be corrupt and the plague shortly to follow.

After the emergence of probability, Petty had a totally different conception of a sign of corruption. In 1671, when campaigning for a central statistical office (some 80 years after Kellwaye's tract) he thought we should determine the life expectation in various communities; this 'Scale of Salubrity' would be 'a better judge of airs than the conjectural notions we commonly read and talk of' [1927, I, p. 171; cf. 1674, p. 87]. Petty was himself a physician. He did not have a better theory about the plague than Kellwaye. Both thought it is consequent upon corrupt air. That is no defect in their science: Even at the time of the Cairo plague of 1834–5 the question of miasmic *versus* infection theories of pestilence remained unsettled. Most people had always blended the two. Thus Danish Bishop Aarhus, in a work translated into English in 1480, said that the 'reek and smoke of such sores is venemous and corrupteth the air' and advised us to flee from sick people to avoid the miasma they caused. 'Fly far and return late' as Thomas Lodge summed it up in his 1603 *Treatise of the Plague* [Ch. 4]. We now find bizarre the theory of Kellwaye that flocks of children corrupt the air, especially at burials – a theory which led to banning poor children from attending the interment of richer folk, doubtless in fact for their own good. Equally useful was Lodge's warning that when 'rats, moles and other creatures (accustomed to live underground) forsake their holes and habitation, it is a token of corruption' [Ch. 3]. This doctrine, which goes back at least to Fracastoro's theory of probable signs, led to sound practice.

A Petty or a Graunt had the same beliefs about plague as a Kellwaye or a Lodge. Their miasmic theory on the origin of plague fit the facts fairly well. In due course it could explain the fact that plague often starts at the dock: foul air had been brought by ships from overseas. And the theory had practical consequences of which we still approve: Flee, avoid animals, and erect houses for quarantine.

It is not on point of medical theory that we distinguish a Petty from a Kellwaye. It is in terms of how to assess the theory. For Lodge in 1603, swarms of mice are evidence of pestilence to come because such swarming is both sign and cause of corrupt air. Whether or not the evidence *is* evidence is part of the theory of corruption. Whether or not something is a sign is itself part of the theory. There is no independent *epistemological* criterion. Only when epistemological criteria can be grasped independently of the causal theory can probability and the use of statistics emerge. Only then shall we find a Petty inviting us to conduct controlled experiments to discover 'whether of 100 sick of acute diseases who use physicians, as many die in misery as where no art is used, or only chance'. [1927, ii, p. 170]. The relationship between the data obtained by such an enquiry, and hypotheses about the efficacy of doctors, is not a causal one – it does not depend on any particular theory of medicine. It is an epistemological relationship independent of the particular subject matter. As soon as men have distinguished epistemological from causal concepts of evidence, we can begin reasoning with Graunt,

The contagion of the plagues depends more upon the disposition of the air than upon the effluvia from the bodies of men. Which also we prove by the sudden jumpings which the plague hath made, leaping in one week from 118 to 927, and back again from 993 to 258, and from thence again the very next week to 852.

Only the weather varies so erratically, week by week. If the plague were passed from person to person we could not explain these statistics. The miasma *versus* infection controversy was centuries old. Here a new kind of data is for the first time brought to bear. As Süssmilch said, Graunt was a Columbus.

There has been some dispute as to who the 'Columbus' was. Graunt owned the rights to the *Natural and Political Observations*, but some of Petty's friends thought Petty was the author. D. V. Glass [1963] has the most recent summary of the evidence. Graunt was a successful merchant who wrote little else. When the book first came out he had a good business, but later his business was burnt down, he fell into Catholicism, and he became somewhat withdrawn. Petty on the other hand published extensively and subsequently wrote much in the same vein as Graunt. He became a man of wide reputation as Surveyor of Ireland, massive war profiteer and Professor of Anatomy and of 'Music' (i.e. the Arts in general). Petty was a man who wanted to put statistics to the service of the state. He made plain their significance for enumerating potential soldiers and for collecting taxes. He had made himself rich by this knowledge by exploiting the defeated Irish, and he thereby saw the real importance of collecting statistics for testing a wide range of hypotheses, even the one about medical efficacy. There remains, however, no good reason to think Petty wrote Graunt's book. We accept Charles II's comment to the Royal Society which was considering Graunt's admission; he told them that 'if they found any more such trades-men they should be sure to admit them all without any more ado'.

Once it became possible for a Graunt or a Petty to look at the data as data, and not as a 'signature' of the plague, it was possible to draw a great many inferences. The bills of mortality listed the number of children christened each week, and classified deaths according to disease. Inevitably the tallies were spotty. In particular the quality of the 'searcher' making the records varied from parish to parish, and this meant that the same disease would be differently diagnosed in different parts of town. Christenings did not truly gauge births because non-conformists and Catholics would not be christened in the established church. All the same there is no holding back Graunt's inventive mind. The course of various diseases across the decades, the number of inhabitants, the ratio of males to females, the proportion of people dying at several ages, the number of men fit to bear arms, the emigration from city to country in times of fever, the influence of the plague upon birth rates, and the projec-ted growth of London: all these subjects are examined with gusto.

Practical consequences of this enquiry abound. For example Graunt recommends a guaranteed annual wage. He reasons as follows: (i) London is teeming with beggars. (ii) Hardly anyone dies of starvation. (iii) Therefore the national wealth already feeds them. (iv) They should not be put to work, for their produce will be shoddy and the Dutch (who at Ypres already subsidize idlers) will gain British trade. (v) So, at no extra cost to the Nation we should feed them and keep them from defiling our thoroughfares by begging. Graunt's book came out in January 1662, some months before the passing of the fundamental statute of English poor law, the law concerning 'Settlement and Removal'. The disastrous experience of British 'workhouses' confirmed Graunt's gloomy foreboding (iv). What is notable is not Graunt's thesis. It was much in the air at the time and had been advocated at least thirty years earlier. Only his mode of argument from (i) is new. It had little effect, whereas Paris, without benefit of statistics, in the single year 1654 managed to confine 1% of the population to hospitals for the poor, maimed, and mad.

One signal inference of Graunt's is the first reasoned estimate of the population of London. We know the number of births from the Bills. We have a rough idea of the fertility of women. Hence we can infer the number of women of child-bearing age. From this we form a shrewd guess at the total number of families. We also guess the mean size of a family, and thereby estimate the whole population. Of course the method is crude, and aside from its internal defects, there are other sources of error. Graunt tentatively allows for the effect of the plague not only through its ravages but also because of the exodus from the city, in time of plague, by all those who can afford to escape the corrupt air by moving to the country. Graunt checks his estimate of the population by two other methods of inference. One, briefly described, is a straightforward sampling of three parishes. The other is based on inhabited area and a guess at the density of habitation. His ingenious arithmetic refutes a view current at the time, that London could boast two million souls, but carries little conviction for Graunt's own estimate of 384,000. The first method, which had been confirmed by the other two, was exceptionally fruitful, for unlike survey sampling it could be applied to the past as well as to the present and so Graunt could plot the astonishing growth of the city and also prove that much of the increase was due to immigration, not procreation. He could also show that despite the horrors of plague, the decrease in population of the worst epidemic was always made good within two years.

Graunt's book appeared in the same year as the Port Royal *Logic*. We quoted the *Logic* telling people that if the million of lightning: Of two million persons killed, only one is killed by thunder, and we can indeed say that there is scarcely any violent death less common.' That figure of two million is sheer rhetoric. Arnauld had no exact idea of the relative frequency of death by lightning; he knew only that it is rare. Graunt was better placed:

Whereas many persons live in great fear and apprehension of some of the more formidable and notorious diseases following; I shall only set down how many died of each: that the respective numbers, being compared with the total 229 250 [the mortality over twenty years], those persons may the better understand the hazard they are in [II. 9].

Then follows a page listing calamities and their proportional occurrence. Whereas the *Logic* goes on to speak of the 'probability' of occurrence Graunt does not. The word 'hazard' is a name taken from dicing, but which by 1662 has come to mean peril or danger; it is unclear whether Graunt is using a tired and forgotten metaphor or whether he is conscious of the comparison to games of chance. Graunt has quite a good sense for betting. Here is an example:

Considering that it is esteemed an even lay, whether any man lives ten years longer, I supposed it was the same, that one of any ten might die within one year [XI. 3].

This does not even sound correct, but it is, thanks to the happy choice of figures. Considering that no one had contemplated this problem before, and that Graunt was no mathematician, the solution is not trivial. As he told us in his preface, he gives us 'succinct paragraphs, without any long series of multiloquious deductions', so we have to reconstruct his reasoning. Graunt assumes a uniform death rate, that is, that there is a constant p of dying in a given year. If the chance of living ten years is 0.5, consider a population of size N. The number who survive the first year is $N(1 - p)$. The number who survive the second is $[N(1 - p) - pN(1 - p)]$ or $N(1 - p)^2$. The number who survive ten years is $N(1 - p)^{10} = 0.5N$. Now let q be the chance that *at least* one man in a group of ten dies in a given year; then $1 - q$ is the chance that no one dies. This is just $(1 - p)^{10}$, which, solving the above equation, is 0.5. So, as Graunt says, q is also 0.5.

The correctness of Graunt's observation bears on a matter that has generated some controversy. Although the bills of mortality told Graunt the causes of death and the sex of the victim they did not tell him the ages of death. To get some sort of mortality curve he divides diseases according to whether or not they affect children. He notes the proportion of people that die of children's diseases, and adds half the people who die of such afflictions as measles and smallpox. He concludes that out of 100 people, 36 die by the age of 6. His only other information is that hardly anyone lives to 75, say. According to the searchers, 7% of people die of 'age', but Graunt pays little attention to this. Now he wonders how many people die in each decade. He tells us his answer, but not how he obtained the answer. The famous table is:

Age	Survivors
0	100
6	64
16	40
26	25
36	16
46	10
56	6
66	3
76	1

A surprising number of papers enquire how Graunt might have got these figures [cf. Hull in Petty 1899, Ptoukha 1937, Greenwood 1941, Glass 1963], and it is generally supposed that either Graunt guessed, or else that he applied a rule of thumb. In fact Graunt's table results from first of all solving the equation $64(1 - p)^7 = 1$, and then rounding off to the nearest integer ('for men do not die in exact proportions nor in fractions'). That is, we assume 64 people alive at age 6, and only one at age 76, and solve for a constant chance p of dying in a decade. The solution for p is very nearly 3/8.

Graunt's original readers had no difficulty with this table. I have already mentioned that the Huygens brothers corresponded about annuities, and used Graunt's figures. Christian checked them by working out his own curve of mortality under the supposition of a uniform death rate, and used this for solving problems in joint annuities. K.-R. Biermann [1955b] has found Leibniz following the same reasoning to get the death rate per year rather than per decade. Leibniz's notes on the problem lie together with several others that have to do with dicing. He indicates that they were proposed to him by Roannez about the end of 1675 or the beginning of 1676. By that time, it seems, problems about dicing and about mortality rate had been subsumed under one problem area. Leibniz first toys with the solution (which seems to have been put to him by someone) that if 36 people in 64 die in ten years, then on average 3.6 people will die each year. He then points out that this result leads to a contradiction, and that the mortality curve must be logarithmic. Characteristically his pages are less tidy than those of Huygens, and he has got the figures wrong – it was 36 who died *before* the age of 6, leaving 64 survivors, rather than a population of 64 with 36 fatalities in the next decade. Despite his sloppiness, he, Huygens and Graunt employ the same principles.

The assumption of a uniform death rate after the age of 6 will strike most modern readers as wild. In the next chapter we shall have occasion to examine more ample statistics and show that in fact the assumption was reasonable. Graunt, however, could not know this. To fill in this and other gaps Petty hoped that the nation might found a central statistical office to gather data for the whole kingdom [1927, i, p. 171]. He clearly perceived the direct advantages such knowledge would yield.

The number of people that are of every year old from one to 100, and the number of them that die at every years age, do show to how many years value the life of any person of any age is equivalent, and consequently makes a par between the value of estates for life and for years [*ibid.* p. 193].

An 'estate for life' is a life annuity; an 'estate for years' is a sum paid annually for a set number of years. The latter value can be worked out using laborious but known procedures for compound interest. Thus Petty sees the statistical office furnishing the nation with an equitable system of annuities.

Petty was an ingenious man who did not merely moan for lack of good statistics. 'I have had only a common knife and a clout, instead of the many more helps which such a work requires', as he said in his Preface to *The Political Anatomy of Ireland*. He did an analysis of one parish partly to determine the expected age of life. In a discourse on *Duplicate Proportion* read to the Royal Society in 1674, he asserts that 16 is the expected age. Greenwood says this must be a blunder, for the true answer must have been about 28. Not so: we have already seen, thanks to Huygens, that Graunt's table gives an expected age of a little over 18. If in Petty's parish the child mortality rate was only a little higher than Graunt's guess of 3/8, then 16 would be an expected figure. Hull [1899] says that Mallet's tables for Geneva between 1601 and 1700 suggest that 42% of children die before age 6, so Graunt's data may well be conservative, and Petty's calculation quite correct.

Petty's discourse is intended to apply inverse square laws to a host of phenomena – the velocity of ships, the strength of timbers, the effect of oars and of gunpowder, the distance for sight, smell and the like, and, what concerns us here, 'the lives of men in their duration',

Roots of every number of men's ages under 16 (whose root is 4) compared with the said number 4, doth shew the proportion of the likelihood of such men reaching 70 years of age. As for example, 'tis 4 times more likely, that one of 16 years old should live to 70, than a new-born babe [1674, p. 84].

Note that Petty is cheerfully measuring 'likelihood' by this time, whereas no epistemic word gets measured in Graunt's book. Unfortunately we do not have before us the statistics for Petty's parish. They led Petty to conclude that 'it is five to four, that one of 26 years will die before one of 16; and 6 to 5 that one of 36 will die before one of 26'. Thus Petty rejects Graunt's hypothesis of uniform death rate, supposing that the mortality rate after 16 increases with age. This difference between them was also enacted in correspondence between Hudde and de Witt in Holland. These men did more with their data than Petty ever could, and to them we must now turn.

13

ANNUITIES

Two parties A and B may agree that A pays B a lump sum while B pays A back in annual instalments. If B needs the money and A wants the rent this is called a loan with interest. It is called an annuity when A wants a secured income for an assigned period. Annuities, as opposed to loans, were a standard way to raise public money, partly because it was possible for a government to sell security in exchange for ready cash, and partly because usury was suspect and not a proper business for the state.

Annuities are of several kinds. Perpetual annuities are straightforward loans paying annual interest to the annuitant. A terminal annuity pays an annual sum so fixed that, at the end of the designated term of n years the capital and interest will all be repaid. A life annuity pays a set sum every year of the annuitant's life. A joint annuity on several lives pays until the death of the last survivor. The terminal annuity presents combinatorial problems: how much should one pay to receive a guaranteed £100 for ten years if the rate of interest is 6%? The life annuity adds problems in empirical probability. The fair price for £100 for life must be the same as that for a terminal annuity for n years, where n is the expectation of life. Joint annuities add a further problem in probability mathematics even if it is assumed that the duration of lives is stochastically independent. If we are more realistic and note that, far from independence, the usual joint annuity is a bet on married couples or shipmates, we require yet further statistical data on joint expectation.

Ullpian, the third century Roman jurist, has left one table of annuities. At age 20 you had to pay £30 to get £1 for life while at age 60 this privilege cost £7. Neither price is a bargain. We do not know the mortality rate in the late empire but we can hardly suppose that the state lost money. Major Greenwood [1940] argues persuasively that the maximum rate of £30 for £1 was not determined by actuarial knowledge but by laws preventing excessive usury. Even if he is right, the state at least understood that the cost of annuities should decrease with the age of the annuitant.

Late Rome was a bad place to buy annuities but early Europe looks as if it could have been a buyer's paradise. An English law of 1540 declares that a government annuity is worth 7 years' purchase. The meaning of this is as follows. You pay a certain sum S, and receive an annual return A, for life, such that at the current rate of interest, compounded annually, a loan of S would be completely paid back in 7 years. Apparently this contract was offered independent of age. Indeed, despite fairly steady official commerce in annuities no British government before 1789 appears to have made the cost of an annuity a function of the age of the purchaser.

It does not follow that the actual management of annuity funds was foolish. Annuities were sometimes given as an endowment at birth. We see from Graunt's data that the infant mortality rate was so high that 7 years' purchase is not so absurd as it might seem. Again, the elderly would purchase annuities to sustain them over their declining years. Such sales need not be foolish. What we lack is any professed theory of the relation between age at purchase and annual payments. There was undoubtedly some practical sensitivity to the problem. For example, joint annuities on lives were not uncommon, especially among spouses or friends. The 1540 act made a joint annuity on two lives worth 14 years' purchase and three lives cost £21. Apparently this was at odds with returns, but only slightly. A rather standard set of tables was prepared by Mabbut in 1686, with the imprimatur of Isaac Newton. Mabbut says that more reasonable joint annuities make two lives cost 13 years' purchase, while three cost 19; in general k lives shall be worth $1 + k(n - 1)$, where n is the number of years' purchase for a single annuity. The laws of compound interest and the current rates of mortality combine to make this law reasonable for annuities on the lives of persons of middle age.

If we disregard the effect of inflation, two factors determine the fair rate for annuities: the mortality curve for that part of the population that buys annuities, and the going rate of interest on long term loans. The first serious attempt to derive judicious prices for annuities was presented by John de Witt to the Estates General of Holland and West Friesland in 1671. Previous practices in the

Netherlands were detrimental to the state. In de Witt's time Dutch towns made two types of contract. Some would sell life annuities at twice the rate of perpetual annuities. Thus if Amsterdam had to pay 6% on loans it offered 12% yearly until death. Alternatively, as de Witt tells us, life annuities were sold at six years' purchase, then at seven and eight; in the majority of all life annuities current in 1671, they were obtained at nine years' purchase. De Witt shows that at the rates of interest then prevalent it would be to the advantage of a healthy person to buy an annuity at even 16 years' purchase.

We shall presently examine de Witt's reasoning. First it is instructive to run over the disastrous history of annuities in Britain. In 1692 the nation tried to raise a million pounds by contracting annuities at 14 years' purchase. The contract was independent of the age of the purchaser. Now this is not necessarily an absurd bill of sale, for perhaps young people, to whom it is so advantageous, do not think of their old age, whereas dotards who will die in a fortnight are attracted to such a contract. But at any rate the terms are 'unfair'. Edmond Halley [1693] the Astronomer Royal, was aided by Leibniz in extracting adequate mortality data from a German town in order to work out the value of single and joint annuities, but no one who was actually selling annuities seems to have paid much attention. Halley's ideas were taken up a century later by De Moivre. The actual mathematics of annuities attracted excellent brains such as Thomas Simpson and Euler. Moreover, better empirical data were forthcoming from the researches of another type of mind. A. Deparcieux made a study of Paris tontines. A tontine is a society every member of which contributes some amount of money or goods; at any future time the total value of the tontine is owned by the survivors, and the last survivor can do what he wants with the total. No government selling annuities made use of the mortality data this implied. In 1741 J. P. Süssmilch published *Die göttliche Ordnung*, which some writers judge to be the first substantial work of demography, albeit motivated by a desire to demonstrate divine providence in the distribution of births and deaths. The facts which he discovered were little used.

Perhaps the first statistical results to be taken seriously were the Northampton tables of 1780, devised by Richard Price, the moral philosopher who communicated Thomas Bayes' famous essay on probabilities to the Royal Society. He worked from parish registers in Northampton, and produced tables that became the usual standard of British and American insurance companies for nearly a century. The Institute of Actuaries did not do anything much better until 1869. But even with relatively accurate mortality curves the abuses did not end. Price's tables were rightly conservative for insurance, that is, they provide a statistical cushion for the insurer. The annuitant bets opposite to an insured person: he hopes that he will live long, and so parts with a lot of money now, while the insuree fears that he will have but a brief span, and so parts with some small premium a year in exchange for a lump sum when he dies. Thus tables conservative for insurance are radical for annuities. In 1808 the British government, hard-pressed by war and inflation, decided to issue annuities 'soundly based' on Price's tables. Hence it lost millions of pounds because people live longer than is implied by any sound insurance table.

Naturally anyone who understood what was going on could profit handsomely by this record of bureaucratic incompetence. Conversely, if the bureaucrats understood matters, they could well serve the state:

There is a general persuasion that the life annuity upon two lives, at 17 years' purchase, is much more advantageous than upon one life, at 14 years' purchase [...] it may even be that the joint annuity if sold at 18 years' purchase would be preferred to that upon a single life at 14 years' purchase; as this might produce a notable advantage to the republic, it is, in my opinion, of the highest importance to leave people in this persuasion [cf. Hendriks, 1853, p. 101].

This is what de Witt wrote to Hudde on 2 August 1671. Evidently he was quite ready for the republic to benefit from the ignorance of its citizens. He was then at his apogee, as chief statesman of the Netherlands, briefly the most rational nation of Europe. His correspondent, John Hudde, was Mayor of Amsterdam. Both had made substantial contributions to the Cartesian mathematics. De Witt so successfully extended the theory of conic sections that he was called a second Euclid. Hudde prepared the way for the differential calculus of curves. Both were citizens in the best sense; Hudde, for example, applied his mathematics to determining the slope of dykes and the curve of mortality. By comparison Graunt and Petty are gifted dilettantes. The Dutchmen were well placed. Not only had they more mathematical ability than the English but also they knew the work of their countryman, Huygens, and were actually able to consult him from time to time. As administrators

they had direct access to annuity records and had what they hoped
was an opportunity to have practical effect.

De Witt presented his advice to the Estates General of Holland
and West Friesland in 1671, so we are well informed as to his
method of reasoning. We also possess part of a correspondence
between Hudde and de Witt. They did not altogether agree on how
to analyse the data. We have more of de Witt's argument than of
Hudde's, but we do possess Hudde's mortality table, in a version he
sent to Huygens, and which we presume he sent to de Witt as well.

The basic problem confronting these men is to determine how the
republic should sell life and joint annuities. Dutch towns sold and
paid annuities in half-year units. Hence if we know the half-year
mortality rate, and the rate of interest, we can, by simple but utterly
tedious computation, work out how much an annuity should cost.
Much of de Witt's monograph is a compendium of the calculations,
(vouched for as accurate, line by line, by Hudde, and by two
accountants). But he also states his method of reasoning, in strict
accordance with Huygens' textbook. Just like Huygens and Spinoza
and others, he must first demonstrate that mathematical expectation is a valid concept. This he does in a sequence of propositions.

The first proposition is taken from Huygens. De Witt invites us to
imagine that there is some object which determines device which gives
us equal chances, such as drawing lots, 'by odd or even, head or tail,
blank and prize'. He then argues that two chances, one worth
nothing and one worth £20 are exchangeable for a certainty of £10.
Huygens is neatly paraphrased. Next de Witt argues for a uniform
mortality curve 'limited to the time when a man is in full vigour'.
Full vigour spans from the 3rd or 4th year to the 53rd or 54th. De
Witt's argument is circumspect. In the second proposition he
concerns himself with a single year which because of Dutch practice
with annuities must be regarded as two half-years.

It is not more likely that the man should die in the second half year of the
aforesaid year than in the first [. . .] One finds an equality of chance similar
to the case of a tossed penny, where there is an absolute equality of
likelihood or chance that it will fall head or tail, although it depends entirely
upon chance as to the side on which it shall fall, and this to so high a degree
that the penny may fall head 10, 20 or more times following without once
falling tail, or vice versa.

De Witt therefore thinks that the likelihood of death (as I have
translated the Dutch *apparenz*) in a given half year, given that one

115

dies in one or the other half, is like the equal chance (*kans*) with a coin which must fall head or tail. It is of no significance if we find in some tables that some years have most of the deaths in the second half rather than the first – no more of significance than if in coin tossing we get a long run of tails. This observation must have been to ward off potential criticism based on actual tables, for Hudde's empirical record of annuities, to be described below, does show some striking aberrations. For example there are vastly more deaths at age 36 than at 35, or at 70 than 69, and de Witt, with no theory of curve fitting, has to use rhetoric to make us ignore such curiosities.

So far de Witt argues only within a single year. To extend the reasoning to Proposition III, consider any pair of years composed of the half years *a, b, c* and *d*. Now *a* and *b* are by Proposition II 'equally deadly'. But *b, c* constitutes a year's duration. So we can use Proposition II to infer that *b* and *c* are equally deadly, and hence that (*a, b*) is as deadly as (*c, d*). The argument seems specious. Biermann and Faak [1959] show that Leibniz, in his notes on de Witt's book, is unsatisfied. Proposition III appears to commit the fallacy of the heap. (A heap of stones is still a heap if you remove one stone, but although there is no particular stone-removal that changes a heap to a non-heap, removing 300 stones can turn a heap into a non-heap.) If *a* and *b* are the two halves of year 4, and *y* and *z* are the two halves of year 54, then *a* and *b* may be pretty well equally deadly, and so may *y* and *z*. Moreover, there may be no pair of successive half-years *m* and *n* which are not more or less equally deadly. Yet *a* and *z* may still have very different chances of mortality. However I believe that de Witt is not really trying to prove that *a* and *z* are equal. He has a different motive. His audience will have had only one kind of experience with long aberrant runs from purely random devices. They will know that coins and dice are sometimes freakish but no one ever experienced repeated sampling from a lottery with one hundred tickets, and so no one has observed that, by chance, ticket 36 may be drawn far more often than ticket 35. So de Witt has to urge that the lottery may combine irregularity within the random. We must not expect every year to have about the same number of deaths; we are to demand only that there is no discernible trend.

De Witt wishes to compare his argument as closely as possible to a lottery. But there is no Fundamental Probability Set of equally probable tickets around. De Witt is rightly unperturbed:

116

Annuities

'It is not the number of chances of each value that we must consider, in the application of the aforesaid rules, but solely the reciprocal proportion.'

He proceeds to assign unit chances to death over the first 100 half years, then 2/3 for the next 10 (from age 55 to 60) then 1/2 for the next ten half years, and 1/3 for the rest, killing off everyone by age 77, which, given the rare chance of survival into the 80s is, he says, about right.

Hudde appears to have favoured an alternative model, with a uniform death rate up to age 81. He went through the records of annuities sold by the United Provinces in 1586–90. Since he is doing this work in 1671 every annuitant is at least 80 by the time the tabulation is made. I do not believe there is an extant copy of the table that Hudde sent to de Witt. Presumably the table he sent Huygens is identical. There are fifty columns, the nth column being for persons who took out an annuity at age n. The column lists, for each annuitant, the number of years for which he collected his annuity, so that n, plus this number, gives his lifespan. On the basis of this data de Witt believes himself vindicated·

The equation of mortality from year to year may with confidence be defined as ascending from the age of 50 to that of 75 years, and it is found that, if one chooses several lives, aged 50 years, and without enquiring whether the person be of a good or bad state of health, that is, taken at random as their lives are then found to be – they die almost exactly (at least without any sensible difference) as follows:—From 50 to 55 inclusive, 1/6: from 55 to 60. 1/5 [. . .] From 90 to 100, 1/1.

What de Witt says is correct but requires some qualification. If we use the table Hudde sent to Huygens to select all annuitants who live to be 50, and then determine the proportion of survivors who die every five years, de Witt's figures are, as he says, about right. But we must consider several other factors, which may be illustrated by considering as extremes the column for annuitants who start at age 1, and those who start at age 50. First, the age at entry, and hence at death, will be accurate for persons in the former table, but not for the latter, because people of about 50 cannot be relied on either to know their age or to tell it truthfully. Second, people about 50 who are buying annuities are presumably hale and hearty, whereas people about 50 who have held annuities since age 1 will not, in general, be healthy. Naturally it is difficult to confirm the first effect, but the second effect is easy to check: a great many people who had annuities bought for them when they were under 5 died between 45

117

and 50, whereas no one who bought an annuity after stated age 43 died before he reached stated age 50.

We may have two quite different interests in Hudde's table. First, we may want to construct a mortality curve for that part of the population which is rich enough to buy annuities. For this purpose we will not want to examine the survival rate of relatively late purchasers of annuities. I have examined the data for all those who had annuities purchased before age 16 (i.e. before the end of Graunt's first 'decade'). These indicate a fairly uniform death rate up to about 46, followed by a sharp increase after that date. Thus, qualitatively de Witt is still right. However if we are interested in selling annuities to mature men such a mortality curve is not relevant. We need to know the curve for persons who say they are over 45 and want to buy annuities. According to Hudde's table after a few years these people have an essentially uniform death rate, much less than in the population at large. I infer that when Hudde argues against de Witt, and in favour of a uniform death rate, he has in mind the problem of selling annuities to mature men, rather than the death rate in the whole population. Thus the disagreement between these two able mathematicians concerns a somewhat delicate matter of the purpose for which the data are to be used.

De Witt's memoir was almost lost. There was only one edition of thirty copies. Todhunter knew about it only because it is mentioned in correspondence between Jacques Bernoulli and Leibniz in 1703–4. K.-R. Biermann and M. Faak [1959] have discovered the origin of this correspondence. In 1700 a German review of current literature had a brief notice of a French translation of work by William Temple, edited by Jonathan Swift. This includes some correspondence between Temple and de Witt. The review thought this might be the sole published material of de Witt's. A few months later it issued a correction: Witt had published a book on the geometry of conic sections, and 'a short but insightful little study of the calculation of the value of annuities'. This caught Bernoulli's eye and he asked Leibniz about the book. Leibniz said he could not then find the book amid his masses of papers, but promised to keep an eye out for it. He never did find it.

In describing the book to Bernoulli, Leibniz says that de Witt employs the 'usual method of equally possible cases'. This is not a literal description. 'Equipossibility' does not occur in de Witt; rather, as we shall see in the next chapter, this concept which played such an important role in probability is due to Leibniz. In 1703 he was relying on memory, but at one time he had made a careful study of de Witt's tract. In particular, as noticed above, Leibniz was critical of de Witt's argument for a uniform mortality curve between 4 and 54. It was a topic in which Leibniz dabbled from time to time. In the preceding chapter we saw that in 1676 Leibniz tackled a problem arising from Graunt's mortality table. Stimulated by the appearance in 1682 of Petty's *Essay in Political Arithmetic* he published two papers on the topic. Although he had been critical of de Witt's confidence in uniform mortality curve between 4 and 54, he now has changed his mind, acting on 'the fundamental assumption that 81 newly born infants die uniformly, that is to say, one a year in the 81 following years'.

Louis Couturat, usually supportive of Leibniz, condemns this axiom as 'absolutely gratuitous', and a disastrous instance of *a priorism* [1901, p. 274]. Yet the same paper lists some 56 questions about population statistics which are to be answered empirically, and this indicates more respect for observable facts. Leibniz does say some fanciful things about uniform death rate until 81: for example, the final fatal year is both attested by scripture and is the fourth power of three, and so is a very 'reasonable' lifespan. But although Leibniz is prone to over value harmonizing of *a posteriori* and *a priori* evidence, he is not one to ignore empirical facts. Elsewhere we find him contending *a priori* that no two things in the universe are identical, and in support of this cheerfully challenging a gentleman to find two identical leaves in a garden, or inviting us to find two identical drops of water when viewed under the new microscope. Likewise it is not hard to locate an empirical basis for his *a priori* mortality curve. Leibniz had visited Hudde in November 1676. In January of that year he had written notes on Roannez's problem in mortality. So he had both the occasion and the interest to obtain Hudde's views. Moreover, as he told John Bernoulli, he had discussed these questions with an acquaintance of Hudde's, John Jacob Ferguson, a Scot who briefly worked in Hannover and who had settled and published in the Netherlands [*M.S.* III. ii, p. 767]. Moreover, Leibniz wanted to print Hudde's table among the appendices to a new edition of the *Art of Combinations*. So it is easy to conjecture what must have been in Leibniz's mind. Hudde, he knew, had contended against Witt that annuities should be calculated on the basis of uniform mortality. So he supposed that the empirical data did confirm the hypothesis of uniform mortality. In fact they do not. However, Leibniz may have had before him a memorandum of Hudde's table, or perhaps just the first column, showing the ages at death of all persons who get annuities before age 1. People in that column do, as it happens, die with more uniformity in the middle years than is true on the average. Moreover, the oldest survivor dies at age 81. (Or is this too an artefact of the table? Hudde compiled his list just 81 years after the last of the five years he studies, and it is not clear that there do not exist living survivors.) If Leibniz had surveyed the whole table he would have found people alive until 97, so we guess that he or Ferguson had only a memory of the first column. Yet subsequent work by Halley and De Moivre reveals that this is another instance of Leibniz's infuriating ability to get the right answer by an unjustifiable inference from the wrong data.

In 1692 the British government devised one of its characteristically unwise annuity schemes. There had been some public discussion of how it should go, and this aroused the interest of the Astronomer Royal, Edmond Halley. Since there were no adequate statistics in England, he enlisted the help of a German pastor, Caspar Neumann of Breslau. Leibniz acted as intermediary. On the basis of five years of exact records of ages at death he is able to draw up the best table of mortality so far and then combines this data with his solution to problems in joint annuities [1693]. Halley's table remained a standard for some eighty years, because subsequent information coincided with his estimate of mortality. In particular it is used in De Moivre's classic 1725 textbook on annuities. De Moivre was undoubtedly the finest probabilist of the age. Here is what he says about the mortality curve:

After having thoroughly examined the Tables of Observation, and discovered that Property of the Decrements of Life, I was inclined to compose a Table of Values of Annuities on Lives, by keeping close to the Tables of Observation; which would have been done with Ease, by taking in the whole Extent of Life, several Intervals whether equal or unequal: However, before I undertook the task, I tried what would be the Result, of supposing those Decrements uniform from the age of Twelve; being satisfied that the Excesses arising on one side, would be nearly compensated by the Defects on the other; then comparing my calculation with Dr. *Halley*, I found the Conclusion so little different, that I thought it superfluous to join together several different Rules, in order to compose a single one: I need not take notice that from the Time of Birth to the Age of

Twelve, the Probabilities of life increase, rather than decrease, which is a reason of the apparent Irregularity of the Tables in the beginning [1725, p. ix].

He assigned a probability of death of 1/86 to each year, rather than 1/81. He has more data. Halley's oldest man is 84. Graunt's was thought to be 86, but this was just a consequence of Graunt's computing in decades to the age of 6: it has no empirical significance. De Moivre also used

Tables of Observations made in *Switzerland* about the Beginning of the century, wherein the Limit of Life is placed at 86. As for what is alleged, that by some Observations of late Years, it appears, that Life is carried to 90, 95, and even to 100 years. I am no more moved by it, than by the Examples of *Parr*, or *Jenkins*, the first of which lived 152 years, and the other 167 [*Ibid.*, p. x].

To sum up: Graunt devised a mortality table on the simple assumption of uniform mortality. Petty tried to do better. De Witt imagined that we have uniform mortality in the prime of life but that the death rate increases after age 54. Hudde contended that for computing annuities we should use uniform mortality. Leibniz once criticized uniformity but later accepted it. Halley's table points against uniformity; De Moivre shows Halley's curve can usefully be flattened. All this has some significance to the conceptualization of probability. If we consider only fair games it is natural to have a theory founded on a set of equally likely alternatives. but as soon as we get real life statistics, we expect the model to be inadequate. However, the statistics actually available chiefly concern the mortality curve. Far from displacing the model of equal chances they actually confirm it. Thus by what one is tempted to call an accident, both *a priori* and *a posteriori* considerations made the model of a Fundamental Probability Set of equal chance definitive of probability. Without much distorting the facts Leibniz is able to recall, when writing to Jacques Bernoulli, that even de Witt conducted his reasoning by the method of 'equally possible cases'.

14

EQUIPOSSIBILITY

In 1703 Leibniz told Bernoulli that Witt had computed annuities by the 'usual method of equally possible cases'. Leibniz was wrong about de Witt, but his remark shows he was familiar with equipossibility. It is commonly supposed that this concept originated with Laplace around the end of the eighteenth century, but in fact it was commonplace at the beginning. Laplace did define probability as the ratio of favourable cases to the total number of equally possible cases, but so did Leibniz in 1678. The definition was in full vigour a century after Laplace and is still not dead. Here is an historical problem. How could so monstrous a definition have been so viable? Its inadequacy seems evident to us. I could quote any of a score of eminent critics. Here, for example, is Hans Reichenbach discussing a 'principle of indifference' in the foundation of probability:

Some authors present the argument in a disguise provided by the concept of equipossibility: cases that satisfy the principle of 'no reason to the contrary' are said to be equipossible and therefore equiprobable. This addition certainly does not improve the argument, even if it originates with a mathematician as eminent as Laplace, since it obviously represents a vicious circle. Equipossible is equivalent to equiprobable [1949, p. 353].

Even workers who in our century have defended equipossibility have done so because they have philosophical views about the impossibility of producing non-circular definitions. Thus Émile Borel, to whom all probabilists owe so much, maintained that such circles were not vicious. It is an error of logicians, he thought, to try to produce a non-circular definition of probability [1909, p. 16].

Neither critics in the style of Reichenbach nor defenders like Borel explain why so dubious a concept as equipossibility should have had such a successful career in well over two centuries of lively theorizing. Yet an explanation does exist. It arises first of all from the essential duality of probability, which is both epistemic and aleatory. Aleatory probabilities have to do with the physical state of coins or mortal humans. Epistemic probabilities concern our knowledge. We have seen that the word 'probability' was annexed to this pair of concepts only after 1662. However, a similar duality was already well established for the word 'possible'. Hence in virtue of its ambiguities 'possibility' could usefully define an unclear concept of probability. The definition was potentially equivocal but this was a positive merit. Probability was evolving as something jointly physical and epistemological. Possibility was already dual in a similar (though not identical) way. The definition of probability in terms of possibility is not an historical freak but a fairly essential feature of the development of both concepts.

In modern English the word 'possible' has different connotations in different grammatical constructions. We need not develop this in detail, as is done in my paper 'Possibility' in *The Philosophical Review* for 1967. But notice especially the contrast between 'possible that' and 'possible for'. 'It is possible that John Arbuthnot was joking when he wrote about chance.' That means that for all we can tell he may have been joking. 'It is possible for Arbuthnot to joke about Queen Anne.' That is, Dr Arbuthnot is able to joke about his royal patient. The first possibility is relative to our state of knowledge and has long been called epistemic. The second possibility says it is physically possible for that wag to joke about his sovereign – neither a dull brain nor a cruel monarch prevent him. 'Possible that' tends to be epistemic (unless preceded by the adverb 'logically') while 'possible for' goes with actual abilities independent of our knowledge of them. English has its grammar to mark these distinctions, but the difference is not peculiar to this language. Thus Richard von Mises notices how a comparable ambivalence in the German *möglich* is utilized in equipossibility definitions of probability:

Ordinary language recognizes different degrees of possibility or realizability. An event may be called possible or impossible, but it can also be called quite possible or barely possible (*schwer oder leicht möglich*) according to the amount of effort that must be expended to bring it about. It is only 'barely possible' to write longhand at 40 words per minute; impossible at 120. Nevertheless it is 'quite possible' to do this using a typewriter [. . .] In this sense we call two events equally possible if the same effort is required to produce each of them. This is what Jacques Bernoulli, a forerunner of Laplace, calls *quod pari facilitate mihi obtingeri possit* [. . .] But this is not what Laplace's definition means. We may call an event 'more possible' [*eher möglich*] than another when we wish to express our conjecture about whatever can be expected to happen. There can be no doubt that equipossibility as used in the classical definition of probability is to be understood in this sense, as denoting equally warranted conjectures [1951, p. 78].

Thus according to von Mises the epistemic concept of probability corresponds to an epistemic concept of possibility, while the aleatory concept of probability corresponds to a concept of physical possibility. This is also indicated by a recent article on Bernoulli by P. M. Boudot [1967], who says that these distinctions fit the pattern of the schoolmen who distinguished *de re* and *de dicto* modality. Taking the Latin tag literally, a modality is *de re* if it pertains to things, and *de dicto* if it applies to what is said or can be stated. 'It is possible for Daniel to get to San Francisco by noon', is about Daniel, and thus *de re*. If I say, 'It is possible that he is there now', I appear to say something about the proposition, 'he is there now', and, since a proposition is what is stated, the possibility is *de dicto*. Unfortunately the distinction between *de re* and *de dicto* modality has seldom been drawn with much clarity. Moreover, it has been a different distinction at different times. The scholastics who worked hard at these notions understood them differently from modern logicians. But at a very gross level we may, with Boudot, say that aleatory probability is *de re*, having to do with the physical characteristics of things, while epistemic probability is *de dicto*, for it concerns what we know and hence what can be expressed by propositions. I contend that probability was readily defined in terms of possibility because the *de re*, aleatory, side of probability matched *de re* possibility while the *de dicto*, epistemic side matched *de dicto* possibility. This is a difficult thesis because the *de re/de dicto* distinction in the seventeenth century is not identical either to any present distinction nor to any scholastic distinction. I shall elaborate on this in the next chapter, but for the present let us use the hints of von Mises and Boudot to study equipossibility definitions of probability.

First let us review the terms used to refer to aleatory probability. How did one say that as a matter of physical fact a number of events have equal (aleatory) probability? We have seen Cardano invoking the Aristotelian concept of *potentia*, of the power or ability of an event to happen. Galileo spoke of events happening with equal ease: cognates of the Latin *facile* were regularly employed. Another

124

word which crept into use was 'proclivity'. Interestingly the translator of Daniel Bernoulli's 1730 treatise on utilities translates *proclives* throughout as 'probable'. In most contexts such a translation would be bizarre, but when Daniel says several events have equal proclivity, he means that they have equal aleatory probability. His usage had already been established in the Latin version of Huygens' textbook. In his proposition I, we read *aeque facile* while in proposition III we have *casus aeque in proclivi*, namely events, or chances (translating the Dutch *kans*) that have an equal proclivity to occur. This terminology is used again by Jacques Bernoulli in his 1705 *Art of Conjecturing*, posthumously published in 1713. In the most original part of Jacques' book, namely the Part IV which is the topic of our Chapters 14 and 15, we find him saying, 'All cases are equally possible, that is to say, each can come about as easily as any other' (p. 219; *omnes casus aeque possibiles esse, seu pari facilitate evenire posse*. Cf. the similar quotation used by von Mises above.) Almost immediately after this passage several such cases are called *aeque proclives*. In short, Jacques Bernoulli used facility, proclivity and possibility as near synonyms. Despite the fact that he is justly famous for emphasizing the epistemological side of probability, in these passages he is concerned with its physical side, with events that can be made with equal ease. Possibility is one of the words used to express this idea.

Before the 1713 publication of *Ars conjectandi* no well-circulated work makes use of equipossibility. Bernoulli himself did not employ it much. He got it from Leibniz, who had long associated it with probability. In a memorandum headed *De incerti aestimatione*, dated September 1678, Leibniz asserts that probability is degree of possibility [Biermann and Faak, 1957]. Undoubtedly this *probabilitas est gradus possibilitas* is the ultimate source of the Laplacian definition of probability. The paper in which it appears has the old division problem for its focus, but it aims not at solving the problem (whose solutions are by 1678 well known) but at justifying the solutions. The format is similar to that of Huygens' textbook. The problem is to justify the use of expectation in computing fair stakes for a game. Leibniz had a special interest in this because he had access to Roberval's posthumous papers, and apparently believed that Roberval's objection to Pascal was based on a doubt about expectation. Moreover, he has his own peculiar point of view: the division problem is a part of jurisprudence, which is to determine the rights of the various players. Indeed questions of inheriting chances in a game are touched on. But the chief distinction between this tract and that of Huygens is that it delves more deeply into foundations. Huygens wanted to justify expectation, and did so by the device of equivalent gambles. He took for granted the simplest case of a fair lottery. He did not ask what makes a lottery fair. Leibniz does.

Leibniz's sketch contains two modes of reasoning, or at any rate, he says that the conclusion can be reached in two ways. I think that one of these is epistemic, and one is aleatory. This may be an anachronism; it is too tempting to wish upon Leibniz distinctions clear to us and obscure to him. His first argument employs a version of what in English has come to be called the principle of indifference. This phrase was coined by J. M. Keynes in his *Treatise on Probability*. An older term is 'principle of insufficient reason', used in J. von Kries' probability textbook of 1871. Insufficient reason is certainly an apt name, because Leibniz claims that *his* axiom for probability 'can be proven by metaphysics', and the metaphysics in question is the ubiquitous Leibnizian 'principle of sufficient reason'. Unfortunately this principle of Leibniz's is too pandemic. He used it to investigate the origin of the universe, the nature of truth, the elements of grammar, the basis of optics, the essentials of statics, the laws of dynamics and, here, the foundations of probability. Students of Leibniz have distinguished endless versions of 'principles of sufficient reason' where Leibniz blithely speaks of but one principle. The reader may consult any book on the philosophy of Leibniz to get some idea of the problem of interpretation.

The application of sufficient reason to gaming is easy. If several players engage in the same contest in such a way that no difference can be ascribed to them (except insofar as they win or lose) then each player has exactly the same ground for 'fear or hope'. Huygens gave us the example of the man who holds a coin in one hand or the other, we know not which; Leibniz states the general criterion for indifference. It appears to be an epistemological criterion. This is confirmed by the very title of our document, for *aestimatione* denotes the formation of opinion. Further evidence of epistemic intent is furnished by the use of the word *opinio* throughout Leibniz's first argument (e.g. *opinio autem de futuro eventu spes metusve est*). *Opinio* and *aestimatione* together with 'probability' itself, are the chief categories of the scholastic epistemology familiar to Leibniz. Despite this there is an ambiguity in the expression, 'no difference between the players can be ascribed to them'. It may mean that so far as we know, no relevant difference can be ascribed, or it may mean that no relevant difference exists. But since Leibniz is here preoccupied by *opinio* I guess he intends the former, epistemological, sense.

On the basis of his axiom Leibniz gives an informal argument (to be followed by a more formal one) for using what we now call expectation as the value of a game. This is a somewhat wordy and philosophically tortuous version of Huygens. Then he says we can reach the same conclusion by another route, for 'probability is degree of possibility'. Fortunately he gives some indication of what he means by possibility. He speaks of the power to achieve various events, and uses the phrase later copied by Bernoulli, *aeque faciles seu aeque possibiles;* 'equally easy, that is to say, equally possible'. Very much later, in 1714, he wrote to Bourguet that,

The art of conjecture is founded on what is more or less easy [the French word *facile*] or, to put it better, more or less feasible [*faisable*], for the Latin *facilis* derives from *faciendo*, which is an exact translation of feasible [*P.S.* III, p. 569].

Similarly, in one of his interminable sets of definitions for the perfect logical language, he says, '*facile* is what is very possible, that is to say, for which little is required' in order to bring it into existence [*S.S.* VI. II, p. 496]. Perhaps Leibniz meant different things by *facile* at different times, but I doubt it. Facility, unlike possibility, is not an evolving concept whose meaning changes as Leibniz develops his metaphysics. It is the same 'facility' used by Galileo and other early writers. Possibility is explained in terms of this non-epistemological concept. Leibniz next conveys the Laplacian definition of probability: when cases are equally possible, the probability of a subset occurring is its 'aliquot part' of the whole. Note that (contrary to Reichenbach) this first equipossibility definition is not part of an epistemological principle of indifference or insufficient reason. Leibniz had two justifications of expectation, one based on insufficient reason, and one based on physical equipossibility. He does not actually say there are two distinguishable ideas of probability but he does present two differentiated arguments. To recall Huygens' example of the man with a coin in one of two clasped hands, we may argue from insufficient reason: there is no way to discriminate between the two hands. Or we may argue from causality: the world is such that it is as possible for the left hand to get the coin as for the right.

If we take seriously the notion of objective possibility, feasibility, proclivity, propensity, or whatever we call it, such degrees of feasibility may themselves be objects of knowledge, to be known with varying degrees of precision. Leibniz was sometimes inclined to say that we ought to determine these degrees *a priori*, but Jacques Bernoulli persuaded him that there are occasions when only *a posteriori* observations will inform us. In these cases,

One may still estimate likelihoods [*vraisemblances*] a posteriori, by experience, to which one must have recourse in default of a priori reasons. For example, it is equally likely that a child should be born a boy or a girl, because the number of boys and girls is very nearly equal all over the world. One can say that what happens more or less often is more or less feasible in the present state of things, putting together all considerations that must concur in the production of a fact [*P.S.* III, p. 570].

A whole battery of familiar concepts is now beginning to emerge. First, there is the relative frequency with which different outcomes occur – 'what happens more or less often'. Secondly, there is feasibility or propensity, determined by the 'current state of things', and investigated either *a priori*, i.e. by deduction from a theory, or *a posteriori*, by observing frequencies. Here we are still in the domain of physics. How is this to be applied to the formation of opinion? Leibniz's best answer is, *Quod facile est in re, id probabile est in mente* [*S.S.* VI. II, p. 492]. That is, our judgement of probability 'in the mind' is proportional to what we believe to be the facility or propensity in things. Note that here probability is not degree of physical possibility, but, as Leibniz puts it elsewhere, 'degree of certainty'.

Leibniz was probably confused and he almost certainly vacillated in his conception of probability. He sometimes leans to an epistemic notion, sometimes to an aleatory one. We shall see in the next chapter that for Leibniz the contradiction is not great, or rather, is overcome by his own peculiar synthesis. Before engaging in such philosophy it will be useful to have a brief survey of the fortunes of 'equipossibility' in the eighteenth century. The terminology caught on in France but not in England. This may be a symptom of the fact that the English workers had a problem situation different from the

French, who more readily perceived the philosophical difficulties about probability.

A good landmark is Diderot's *Encyclopédie.* In the volume covering the letter *P*, dated 1765, we find a judicious survey of kinds of probability, but the only measurable probability is founded on the 'equal possibility of several events'. This applies both to gaming and annuities. The author says of equipossibility that it 'is to be employed when we suppose the several cases to be equally possible, and in effect it is only a supposition relative to our bounded knowledge, that we say, for example, that all the points on the die can occur equally' [xiv, p. 396].

This epistemological conception, 'relative to our bounded knowledge', is perhaps as old as Aristotle but it made its first modern appearance in the 1713 *Ars conjectandi* and thereafter remained dominant in France. Despite this, there is no English mention of it, not even in the work of Thomas Bayes, posthumously published in 1763. The English did not need equipossibility because they took a purely aleatory attitude to probability. They did not need to equivocate by defining probability as possibility which may be either *de re* or *de dicto.* The French clearly perceived the need for both sides of probability but could not face up to it, and took refuge in ambiguous 'equipossibility'.

Naturally no classification of 'national' sciences is exact. One Englishman, William Emerson, writing in 1776, distinguishes 'mathematical probability' from some more general epistemic idea of probability. Of the former he gives a good frequentist account:

Although it is impossible to determine with certainty how an event shall happen yet it may be determined mathematical, what likelihood or degree of probability there is for its happening or failing; and this is all that is intended by a calculation, [...] except that there be made an infinite number of repetitions, and then one with another will always bring it to the same thing as the calculation makes it. [1776, p. 1].

For this calculation 'it is supposed that all chances are equal, or made with equal facility' [p. 3]. But then,

The probability or improbability of an event is the judgment we form of it by comparing the number of chances there are for its happening, with the number of chances for its failing.

Thus Emerson, though operating in the standard English doctrine of chances that speaks only of a physical interpretation, is groping for the idea of probability as 'judgment' or credibility.

On the Continent this groping is far more secure and words are coined to distinguish aleatory from epistemic concepts. J. H. Lambert furnishes an instructive example. In his seminal contributions to error theory he uses some established terminology. Thus in the *Photometrie* of 1760 he says that when positive and negative errors are equally possible, they will occur equally frequently [p. 131]. In the *Phänomenologie* of 1764 he has an extremely valuable discussion of equipossibility (*die gleiche Möglichkeit*, § 152), which he claims to be founded on the arrangement of things in the world, each individual event happening according to its own laws but in such a way that, regarding the collection of events as a set (*Menge*), events of each kind occur equally easily (*leicht*). Thus (he claims) the more frequently an event occurs, the more probable it is in itself. With the possible exception of the last *wahrscheinlich* these are aleatory notions. In contrast his 1765 book on error theory is called *Theorie der Zuverlässigkeit der Beobachtungen und Versuche*. *'Zuverlässigkeit'* or 'reliability' is the word here used for the epistemic probability of observations.

In a similar vein Lagrange, writing in the 1770s, did not invent new words but in his work on error theory took different old words for different ideas. *Facilité* is his favourite for the physical propensity to give errors of varying magnitude. From a known law of the facility of error and a given set of observations, he wants to infer the (epistemic) probability that an unknown quantity lies in a given interval. In his usage, *probabilité*, once a synonym for *facilité*, has become an antonym.

The Marquis de Condorcet, who much influenced the philosophy of his young friend Laplace, furnishes another example of this conceptual torture. In a book applying probability to voting theory published 1785 he writes of the 'purely mathematical sense' in which probability is defined in terms of 'equally possible combinations' [p.v.]. This is illustrated by a die such that each face *'puisse arriver ègalement'*. He asserts that the 'mathematical probabilities' computed thus are merely definitional equivalences to the equipossible distributions with which one started. There is, says Condorcet, a 'more extended sense' of probability. He does not so much go on to define a new sense of probability, as to introduce a new concept, the *motif de croire*. He argues that the grounds for belief are in proportion to mathematical probabilities derived from equipossible cases. Thus in Condorcet's opaque discussion, probability, in the strict sense, appears to have to do with the physical interpretation, whereas there is to be a more general sense, best called 'grounds for belief', which fits the epistemological interpretation.

Condorcet's description of statistical inference is confused, but in a rather engaging way. He is less concerned with inferring the probability of statistical hypotheses, than with guessing what will come next. Thus if we do not know the probability distribution for a chance set-up, but do have some results of trials on it, what is the probability of getting outcome *S* on the next trial? Condorcet's analysis is Bayesian. He says that we conclude not with the 'true probability' of *S*, but with the 'mean probability' [p. lxxxvi].

In a subject prone to conceptual difficulty, I have encountered no phrase less felicitous than '*probabilité moyenne*'. The word 'mean' must be suggested by our starting with a uniform prior distribution, and then averaging. Evidently the 'true probability' of which Condorcet speaks is an objective, physical, unknown, while the mean probability is an epistemological measure of credibility. Elsewhere, Condorcet abandons his attempt to give us distinctions, and in Bayesian analysis speaks regularly of the 'probability of a probability' [p. 180]. The first 'probability' is epistemic, and the second is aleatory.

This then is the background against which we are to understand Laplace's famous definition of probability in terms of equally possible cases. Equipossibility was well enough known to make Diderot's encyclopedia. In that work equipossibility was relative to our knowledge. But Lambert, writing in the same year, kept possibility in its old physical sense, determined by laws of individual objects. It is notable that equipossibility is not a feature of Laplace's early work. Indeed his first paper, of 1774, says innocently that probability is defined in terms of a ratio among cases, so long as the cases are equally probable! [*Oeuvres*, VIII, p. 10]. In the third paper, of 1776, this becomes, 'if we see no reason why one case should happen more than the other' [p. 146]. The word 'possibilité' does not occur in the definition although it does occur soon after [p. 149] to denote something like physical probability.

A simple problem well illustrates Laplace's thought at this time. Suppose we have a biased coin, but no information about the direction of bias. Then in one toss *H* and *T* are equally credible. But in two tosses the four outcomes are not equally credible. If (unknown to us) the bias is for *H*, then *HH* is the most probable outcome; if the bias is for *T*, *TT* is the most probable. In our ignorance *HH* and *TT* are more credible than *HT* and *TH*. Laplace was obviously very pleased with this observation, for he makes it repeatedly, and boasts that no one ever thought of it before.

I said that *H* and *T* are equally credible, while *HH* is more credible than *HT*. Laplace says of *H* and *T*:

One regards two events as equally probable when one can see no reason that would make one more probable than the other, because, even though there is an unequal possibility between them, we know not which way, and this uncertainty makes us look on each as if it were as probable as the other [VIII, p. 61].

Notice that in a careful statement like this, Laplace does not say that *H* and *T* are equally probable, but that we regard them as equally probable. The word 'possibility' is kept to indicate that so far as physics is concerned, there is an objective difference between *H* and *T*. Laplace has not yet become confidently subjective. In contrast, let us turn to the more familiar, polished, Laplace whom all of us have read.

The opening prose of Laplace's philosophical essay on probability is almost as captivating as the mathematics of Book II of the *Théorie analytique* it served to introduce. Laplace's demon has become the byword for a physically determinate system. Because the world is determined, Laplace implies, there can be no probabilities in things. Probability fractions arise from our knowledge and from our ignorance. The theory of chances then, 'consists in reducing all events of the same kind to a certain number of equally possible cases, that is to say, those such that we are equally undecided about their existence' [*Oeuvres*, VII, p. viii]. On the first page of Book II, we have,

One has seen in the Introduction that the probability of an event is the ratio of the number of cases that are favourable to it, to the number of possible cases, when there is nothing to make us believe that one case should occur rather than any other, so that these cases are, for us, equally possible [VII, p. 181].

In the introduction 'equally possible' is glossed in an epistemological way, and the Book II turn of phrase, 'for us, equally possible', is consistent with that gloss.

Whenever Laplace is doing direct probabilities – deductions from probability distributions to other probability distributions – he can happily continue with the epistemological interpretation. Or, one may add, any other interpretation, or no interpretation. His results could nowadays be presented quite formally, as pure mathematics, and the interpretation is irrelevant.

So all goes smoothly until Book VI, on the probability of causes. Here we return to the matter of the 1774 papers. One is inferring from observed data to an unknown probability distribution. It is a distribution of causes, and that is conceived as a matter of physics. One wants to find out the true distribution. And what do we read? As more and more experimental data build up concerning simple events, then, he tells us 'their true possibility is known better and better' [VII, p. 370].

True possibility! We are concerned, he says with discovering an as yet unknown degree of possibility. Throughout this chapter he is altogether consistent. He speaks of the probability that the possibility of an event lies in a given interval. This language occurs even in the introduction; even indeed, in the brief allusion to Bayes I mentioned above.

It seems that 'probability of a possibility' occurs only when Laplace is trying to assess what we now call the inductive or epistemological or subjective probability of what we now call an objective statistical hypothesis. In these circumstances, possibility is *de re*, and is a physical characteristic of the set up under investigation.

Thus Laplace himself is equivocal. When he needs a word to refer to an unknown physical characteristic, he picks on 'possibility', using it in the old, *de re* sense. This was the language of his early papers. When he wants to emphasize the epistemological concept which finally captivated him, he uses 'possibility' in what he makes clear is the *de dicto*, epistemological sense. But even in those introductory chapters, the *de dicto* equally possible cases are ones which we know to be equal because we think of them as being *de re* equally possible, that is, equal in physical characteristics. Leibniz's handy notion of equipossibility made it easier to put off the philosophical problems.

15

INDUCTIVE LOGIC

The equipossibility account of probability enabled continental workers to operate simultaneously with epistemic and aleatory concepts of probability. But for Leibniz, who invented the idea, it meant far more. It enabled him to conceive of probability theory as an integral part of his metaphysics and epistemology. One by-product of this is that Leibniz anticipated the philosophical programme of J. M. Keynes, Harold Jeffreys and Rudolf Carnap, which has come to be known as inductive logic. We cannot study this in isolation from the rest of Leibniz's philosophy. C. D. Broad has called Leibniz 'the greatest pure intellect whom we have known'. One aspect of his intellect is its attempt to combine every aspect of human knowledge, action, and speculation into an elaborately structured but totally co-ordinated unity. We can start on the outside, with inductive logic, but we shall inevitably be drawn into more central concerns.

Leibniz thought that the science of probability would become a 'new kind of logic' (*P.S.* v, p. 448), but this idea lay dormant until around 1920 when it was revived by Jeffreys and Keynes. Later, in the 1940s, Carnap set to work with a particular approach which has, in recent years, been taken as the sole format for inductive logic. The tenets of this programme can be set out tersely as follows: First, there is such a thing as non-deductive evidence. That is, there may be good reasons for believing *p* which do not logically entail *p*. Second, 'being a good reason for' is a relation between propositions. Third, this relation is to be characterized by a relation between sentences in a suitably formalized language. Fourth, there is an ordering of reasons from good to bad, and indeed a measure of the degree to which *r* is a reason for *p*. Fifth, this measure is autonomous, and independent of anyone's opinion: it is an objective measure of the extent to which *r* is a reason for *p*. Sixth, this measure is global – it applies to any pair of propositions (r, p) whatsoever, and not just to some classes of propositions. 'We can always estimate which event under given circumstances can be expected with the highest probability' [*P. S.* vii, p. 188]. Moreover, this global relation is a 'formal' one; it depends solely on the form of the relevant sentences, and not on their content.

It is a matter of some curiosity that no one else was rash enough to emulate these theses until recently, and yet Leibniz believed in them almost as a matter of course. He had grand expectations of his programme. It was part of a scheme for a logical syntax which he called the 'Universal Characteristic'. He prophesied that when it was complete men who disagree would pick up pencils exclaiming 'Let us calculate!' and thereby end contention. Most readers of Leibniz have taken this to be the cry of some alien rationalism which assumes that every issue can be settled by deductive proof. Quite the contrary. Leibniz was not in general speaking of proving propositions but only of finding out which are most probable *ex datis* [*P.S.* vii, p. 200–1].

The 'new kind of logic' has plain enough beginnings. One feature of inductive logic, on which Jeffreys *et al* have insisted so vehemently, is that inductive probabilities are relative to the evidence. We cannot speak of the probability of p; we can refer only to the probability of p relative to r. Leibniz was the only man of his time regularly to declare the relational character of probability judgements. This is a natural consequence of his starting point, namely the law. All legal judgement is *ex datis*, and, as we have seen in Chapter 10, legal reasoning remained Leibniz's paradigm. The new kind of logic was a 'natural jurisprudence'.

A second consequence of this legal point of origin is Leibniz's faith in the objectivity of the probability relation. There may be different assessments of legal evidence, but there is only one *right* assessment. The same goes for probability. Whether or not r is a reason for p is not a matter of personal opinion. For example, most people thought Copernicus was wrong when he promulgated his heliocentric theory. They thought the available data confuted this unorthodox doctrine. Yet, according to Leibniz, it was still 'incomparably more probable' that Copernicus was right [*N.E.* iv, ii. 14].

With jurisprudence for his model Leibniz thought of probability as relational and objective. This creates the notion of an inductive logic but the next steps are harder. Let us suppose our universe of discourse can be represented by some set of disjoint alternatives (for example, the 2^{100} possible outcomes of tossing a penny 100 times). Let t be the empty tautological 'evidence' stating that one of these alternatives will come to pass. If we can assign 'prior probabilities' to the set of alternatives, on the basis of data t, then we can work out the 'posterior probability' for any event (say, the event 100th toss gives heads) relative to any data in the universe of discourse (say, 'first 3 tosses gave tails'). In general, an inductive logician can compute the probability of p, given r, where p and r are in some universe of discourse U, if he knows the prior probability, on tautological 'evidence', of every possible state of affairs expressed in U. It has been an aim of Jeffreys and of Carnap to develop such prior distributions for various U. Although Jeffreys sticks to realistic U and Carnap to simplistic models, they are both moved by two desiderata. One is pragmatic: the resulting posterior probabilities ought, in many cases, to coincide with our unformalized hunches about what is a good reason. The second kind of requirement is 'formal', and consists of conditions of symmetry. For example, suppose that $r(a)$ and $r(b)$ are propositions saying the same thing about individuals a and b respectively; likewise for $p(a)$ and $p(b)$. Then $r(a)$ should be exactly as good a reason for $p(a)$ as $r(b)$ is for $p(b)$. In the case of prior probabilities relative to tautological 'evidence', a set of possibilities with the same formal structure should be assigned the same prior probability. 'Equal suppositions deserve equal consideration': That, said Leibniz, is the first maxim of probability theory, known even to peasant farmers dividing land [*N.E.* IV, xiv. 9].

Unfortunately it is by no means clear which suppositions are equal. Compare the case of three dice, which first cropped up in Galileo's memorandum. Is the supposition that we get two aces and a deuce 'equal' to the supposition that we get an ace, a deuce and a six? The answer is 'yes' if partitions constitute equal suppositions, but not if permutations within partitions are 'equal'. Now as we noted in Chapter 6, it is a matter of empirical fact that dice, photons and electrons obey, respectively, Maxwell–Boltzmann, Bose–Einstein and Fermi–Dirac statistics, each of which assigns different prior probabilities. In the domain of aleatory probabilities like these it is clear how we are to choose the correct prior: by experiment. But how to choose the right priors for inductive logic? In the simplest case Carnap favoured a probability function c^*

based on the Bose–Einstein statistics for possibilities. It assigns equal probability to each partition. He also set out a continuum of priors, excluding only Maxwell–Boltzmann priors for possibilities because of one pragmatic desideratum – the Maxwell–Boltzmann statistics for possibilities preclude learning from experience. The probability of *p* on any non-entailing evidence *r* is just what it was on the tautology *t*. No prior probability in Carnap's 'continuum of inductive methods' carries any conviction whatsoever. In the case of photons and dice, we can discover prior aleatory probabilities – which are prior propensities – on the basis of experiment. But the programme of inductive logic is supposed to be logical, not only independent of experiment, but the very judge of experiment. How can its prior probabilities be chosen? In modern terms there is no good answer. But Leibniz's metaphysics does provide an answer of sorts.

We have distinguished, on the one hand, the epistemic probability that some possibility is realized, and the aleatory or physical propensity for some possibility to exist. In modern opinion the latter exists, so far as we know, only for a fairly small range of chance set-ups. But in Leibniz's opinion every possibility has a propensity to exist. This is a quite specific and central element of his thought. Many people have heard of the Leibnizian doctrine that this is 'the best of all possible worlds'. Less familiar is his dedicated attempt to explain why there is a world at all, and in particular just this one. According to Leibniz every possible world has some tendency to exist, and our world is the one with the greatest propensity. It is important to compare the familiar aleatory concept of 'ease of making' an outcome with the seemingly different concept of making of a possible world, both summed up in our quotation on page 127 above, by a pun on the word 'feasible'.

As we noted in the last chapter, Leibniz used 'equally possible' to mean something like 'having an equal propensity to occur'. He is also the first to use 'possibility' for a quite different idea which, for today's logician, has superseded every other. 'Possible', according to him, means internally consistent. Leibniz was the first modern to understand that proof is a formal matter, attaching to the form of sentences, not to their content. He defined 'necessarily true' as provable from identities in a finite number of steps. Possibility, then, is freedom from contradiction. For us it follows at once that there are two quite distinct concepts of possibility in question; for consistency is not propensity to occur. But in Leibniz's lists of definitions he regularly explains possibility as freedom from contradiction, and in the next breath speaks of one thing being more possible than another. Something is easy or makeable if it is 'very possible'.

Leibniz's twinned notions of possibility are a deliberate part of his metaphysics. His theory about creation, or what he once called 'the radical origination of things' involves possible objects striving for existence. In the esoteric writings we are invited to contemplate consistent notions having more than mere internal consistency: they have a positive drive to come into being. The more they have of this, the more possible they are. 'The possible demands existence by its very nature, in proportion to its possibility, that is to say, its degree of essence' [*P.S.* VII, p. 194]. In emphasizing this aspect of Leibniz's thought, I admit that I am playing down his other story, which is far better known, in which reality is determined by 'the principle of contingency, or of the existence of things, i.e. of what is or appears the best among several equally possible things' [*P.S.* IV, p. 438].

Leibniz's methodology of science always mirrors his metaphysics, but nowhere is the isomorphism more striking than in the analysis of probability. His contemporaries and predecessors all employ some terminology of the 'facility' of getting an outcome with a die. That, he reminds us, means 'makeable'. All early writers speak of the ease of making various outcomes, but for Leibniz ease of making goes with possibility, and probability is degree of possibility. To repeat the quotations of the preceding chapter, experiments show what is more or less makeable 'in the current state of the world'. What is facile *in re* corresponds to what is probable *in mente*. Note the parallel to the metaphysics. In the esoteric writings we do not read so much of a God choosing among internally consistent state descriptions that which describes a most perfect world: God's role is to conceive the possibilities. The creatability of the things will correspond to the degree of possibility in the divine mind. Similarly, in our world the objective propensities of different outcomes to occur are the foundation of our mental expectations, the probabilities, which, as Leibniz had said, are degrees of possibility. Even for Leibniz such an intertwining of a special science and deepest metaphysics must seem bizarre; so I was glad to find that Dietrich Mahnke [1925] anticipated my interpretation, and took the probability–possibility–facility–creatability nexus as a final proof of the way that Leibniz linked ontology and physics. Margaret Wilson [1971] has, however, recently contested this point of view.

We now return to our starting point. Leibniz had learned from the law that probability is a relation between hypotheses and evidence. But he learned from the doctrine of chances that probabilities are a matter of physical propensities. Even now no philosopher has satisfactorily combined these two discoveries. Leibniz's combination, although unsatisfactory, is more fascinating than most. On the one hand we have degrees of makeability *in re*, which we may gloss as tendencies to produce stable frequencies. These are the basis of probabilities *in mente*. In particular cases, such a line of thought can be sound. For example if *r* asserts only that in some chance set-up the objective tendency is to produce outcome *E* on repeated trials with stable relative frequency *f*, then the probability of the hypothesis that *E* occurs on the next trial, relative to this data *r*, is surely *f*. Leibniz appears to be inclined to say that this local piece of reasoning has general application. Just as the possible worlds in the mind of God vie with one another for creation, so all the possibilities that we can distinguish will also have some propensity to be actual. We can apply the calculus of probabilities to these possible worlds, using yet another aspect of Leibniz's grand scheme.

Textbooks on probability nowadays often begin with a chapter on combinations and permutations. Inevitably we take Leibniz's youthful *Ars combinatoria* to be in the same line of business. This early monograph on the theory of combinations confirms his claim to have helped advance probability theory. That he had probability theory in mind is proved by internal evidence and also by his proposal to publish Hudde's tables as an appendix. But that is only a small part of the story. The art of combinations was already an established problem area. It was directed not at probability theory but at ideas.

From any vocabulary of ideas we can build other ideas by formal combination of signs. But not any set of ideas will be instructive. One must have the right ideas. Everyone thought that the right ideas would be simple. From an exhaustive set of simple ideas one would generate all possible complex ideas. This is done formally as an operation on signs for the ideas, and this was the point of the art of combinations. Leibniz's immediate predecessors were enraptured with the thought that the world could be understood from a set of signs. Here I have in mind not so much great figures like

Descartes and Spinoza, but a myriad of lesser intellects whose dozens of programmes for universal grammar were motivated by a belief that if only we could uncover in the collective wisdom of mankind a suitable set of ideas, then we would be able to unlock all the secrets of the universe. Universal grammar, which has recently been presented as a key for understanding mind, was in those days a project for understanding all of nature.

Much of Leibniz's intellectual politics is a part of this ferment. His plans for academies and scientific journals intend to co-ordinate knowledge so that we can discover what are the true underlying ideas. Many of his predecessors hoped to uncover an original language preceding Babel. It would encode the true ideas. Leibniz's better plans did not believe in lost innocence but rather in a science and a language that more and more closely correspond to the structure of the universe. His encyclopedia of unified science would collect all present knowledge so we could sift through it for what is fundamental. With the set of ideas that it generated, we could formulate the Universal Characteristic. The art of combinations would enable us to compute all descriptions of possible worlds that could be expressed with that stock of ideas. And the possible worlds so described would all have some propensity to exist.

The Characteristic was supposed to enable us to compute the probabilities of disputed hypotheses relative to the available data; if our Characteristic is founded on simple ideas, then there will be no finitely stateable *a priori* reason that would cause one possible world describable in our language to come to pass rather than another. We thus have a set of alternatives constituting a Fundamental Probability Set to which we can apply a uniform prior probability distribution. The prior distribution is applied not because our set is one among whose alternatives we are ignorant; it is a set such that by metaphysics we know each element has some propensity to exist. Relative to our finite knowledge we may be able to assign only a uniform distribution over possibilities, but we will slowly correct this, and as we learn more our probability assignments will asymptotically tend to a maximum for the real world, i.e. the possibility with the highest actual propensity.

The notion of an asymptotically improving language may sound peculiar to students of Carnap, who writes as if the language, its logic, and its prior probability distribution are fixed. But that is not an essential feature of his theory. Indeed as early as 1932 a fellow member of the Vienna Circle, Friedrich Waismann, was proposing that language and prior probability should be constantly adjusted as we come to know more about the world. And in Carnap's 1952 *Continuum of Inductive Methods* it is argued that different possible worlds are most efficiently studied by different inductive logics. In the case of Leibniz it is also important to emphasize the role of asymptotic improvement. It is crucial to his metaphysics that no finite analysis is ever complete. In particular no humanly practicable language will enable us to write down an exhaustive classification of possible states of affairs. All possibilities that we can delineate will in fact be complex, and stand for a class of simpler possible worlds. All we can do is estimate propensities for this class, and work hard to make both our classification and our estimates better. We are able to apply probability theory here not because of a principle of indifference applied to an *a priori* language, but rather by a metaphysical ascription of propensities to a classification of possibilities based on learning, scholarship and experiment.

Leibniz's new kind of logic is, then, a compound of three disparate elements. First, there is the doctrine of chances. Secondly, there is a theory of possibility. Some parts of it, original with Leibniz, have become the truisms of our logicians, but the parts most pertinent to probability, though the truisms of yesteryear, are now almost wholly repudiated, and those parts which have their chief role in metaphysics are peculiar to Leibniz. Finally, there is the theory of ideas, a final flourish to the intellectual programme of a preceding era.

Ideas, possibility, and chances create a matrix within which inductive logic could be conceived. They leave open technical questions that have recently perplexed inductive logicians. The problem of choice of initial measure function for prior probabilities is present in what Leibniz proposes, but only after Carnap can we understand it properly. Nevertheless, it is pleasant to note that a Leibnizian ought to like Carnap's preferred c^* which assigns equal probability to what Carnap calls 'structure descriptions'. For Leibniz this could be a methodological consequence of the identity of indiscernibles. Structure descriptions are the finest partitions of possibility that produce descriptions of states of affairs that are distinguishable by the predicates available to a monadic language. A uniform prior distribution over structure descriptions is just c^*.

There is an alternative, however. Leibniz had an important theory of what he called 'architectonic' reasoning. There is deductive *a priori* reasoning, and inductive *a posteriori* reasoning, but there is a middle ground of central importance to science. We favour hypotheses for their simplicity and explanatory power, much as the architect of the world might have done in choosing which possibility to create. The paradigm of this kind of reasoning is our preference for the principle of least time over Descartes' explanation of Snell's (purely inductive) law of refraction [*P.S.* I, p. 195, VII, p. 274]. The Leibnizian might wish his prior probability assignments to conform not to c^* but to Harold Jeffreys' 'simplicity postulate'. Jeffreys needs the simplicity postulate in order to get positive probabilities for law-like propositions, but his is a purely epistemological thesis for which only pragmatic reasons have been given. For Leibniz, in contrast, it is one more pleasant consequence of metaphysics. Simplicity of covering laws and variety of phenomena are the twin measures of perfection for possible worlds. Hence laws with those features will have a greater objective tendency to reality than cumbersome or restricted principles. As in our earlier discussion, high probabilities derived from a simplicity postulate are grounded on a metaphysical ascription of propensities.

We should go no further in reconstructing a Leibnizian theory of probability and inductive logic. Its most notable feature is that there is an objectively correct prior distribution of probability for a set of possibilities. The correct distribution is the one that corresponds to the propensity of the possibility to exist – very much as in dice rolling. I doubt that anyone will accept such a Leibnizian foundation for inductive logic. Still, I prefer it to more recent theories of global inductive logic, which have no foundation at all.

16

THE ART OF CONJECTURING

Jacques Bernoulli's *Ars conjectandi* presents the most decisive conceptual innovations in the early history of probability. The author died in 1705. He had been writing the book off and on for twenty years. Although the chief theorem was proved in 1692, he was never satisfied and he never published. The work was finally given to the printer by his nephew Nicholas, and appeared in Basle in 1713. In that year probability came before the public with a brilliant portent of all the things we know about it now: its mathematical profundity, its unbounded practical applications, its squirming duality and its constant invitation for philosophizing. Probability had fully emerged.

The chief mathematical contribution of the book is plain enough: the first limit theorem of probability. This result has rightly been given the seal of A. N. Kolmogorov as being proven by Bernoulli with 'full analytical rigour' [Maistrov, 1974, p. 75]. But what the theorem means is another question. We shall try to find out in the next chapter. First it is worth investigating Bernoulli's own conception of probability. Since we still lack universal agreement on the analysis of probability no one writes dispassionately about the man. He has been fathered with the first subjective conception of probability. Yet Richard von Mises [1951] could cast him as a stalwart frequentist. More recent statisticians such as A. P. Dempster [1966] say he anticipated Jerzy Neyman's approach to inference via confidence intervals. P. M. Boudot [1967] has argued that Bernoulli was a good inductivist and anticipated the theories of Rudolf Carnap. Since each of these approaches to probability is customarily deemed inconsistent with every other, each school claims Bernoulli as its own. The truth of the matter must be that he was, like so many of us, attracted to all these seemingly incompatible ideas and was unsure where to rest his case.

143

Once a research programme is under way the occasional master-piece of permanent value often has three distinct characteristics. It does something almost completely new which, although much in the air at the time, has never before crystallized, but, once written down, sets the direction for all future enquiry. Secondly, it epitom-izes what everyone has known for a long time but has been unable to state succinctly. Thirdly, and much less often noticed, it ends certain possible lines of development which, until that node in history, were perfectly open but now become closed. The first two features of Bernoulli's work are evident. The third feature is this: until 1713 it was not in the least determined that the addition law for probability would be accepted. Bernoulli was the last master to contemplate non-additive probabilities. I owe this observation entirely to a discussion with Glen Schafer, who himself is now developing important non-additive notions related to probability.

The *Ars conjectandi* comes in four parts. The first is an improved version of Huygen's book on games of chance. Bernoulli had a marked gift for giving intuitive explanations of technical concepts. For example he warns us that 'the word "expectation" is not meant here in its usual sense [. . .] we should understand rather the hope of getting the best diminished by the fear of getting the worst. Thus the value of our expectation always signifies something in the middle between the best we can hope for and the worst we can fear.' Bernoulli's ample grasp of the concepts enables him to distinguish, for example, the several senses of the ambiguous problems set by Huygens. Each possible interpretation provides a problem that is now solved with panache. Likewise, when he wishes to make clear that at best the addition law holds only for disjoint events he gives a vivid illustration:

If two persons sentenced to death are ordered to throw dice under the condition that the one who gets the smaller number of points will be executed, while he who gets the larger number will be spared, and both will be spared if the number of points are the same, we find that the expectation of one of them is 7/12 [. . .] it does not follow that the other has an expectation of 5/12, for clearly each has the same chance, so the second man has an expectation of 7/12, which would give the two of them an expectation of 7/6 of life, *i.e.* more than the whole life. The reason is that there is no outcome such that at least one of them is not spared, while there are several in which both are spared.

Everyone knew that the addition law can hold only for disjoint events, but this fact had not before been presented so graphically. Later we shall find examples refuting the addition law even for disjoint events.

In Part II Bernoulli gives a general essay on the theory of combinations. In Part III he applies this to a sequence of further exercises on games of chance. Elegant and generalized as this work is, and novel as are some of its solutions, it is chiefly the application of a powerful mind to a familiar problem area. It is Part IV that revolutionizes probability theory. The revolution is twofold. For the first time a 'subjective' conception of probability is explicitly avowed, and the first limit theorem is proven.

Part IV intends to show the application of probability mathematics to matters of economics, morality and politics. It is this part that justifies the very title, *Ars conjectandi,* which is patterned after *Ars cogitandi,* the Latin title of the Port Royal *Logic.* The art of conjecturing will take over where the art of thinking left off. Bernoulli does not in fact make practical applications to economics or morality. It is of interest to compare the 1708 book on games of chance by Pierre Rémond de Montmort. Like all the mathematical Parisians Montmort had heard about the contents of Bernoulli's book, but had not seen it, and, since Bernoulli had died in 1705, supposed that it would never be published. 'If I were going to follow Bernoulli's plan I should have added a fourth part to apply the methods contained in the first three parts to political, economic, and moral problems. I have been prevented by not knowing where to find theories based on factual information, which would allow me to pursue such researches.' Perhaps Jacques Bernoulli also lacked such facts and such theories, but the thing Montmort most lacked was the limit theorem which would show how observed frequencies are related to underlying chances.

At the outset of Part IV Bernoulli announces that 'Probability is degree of certainty and differs from absolute certainty as the part differs from the whole.' This notion of degree of certainty is not new. It is found in Leibniz's 1678 *De aestimatione,* and Leibniz had been telling the idea to correspondents for a long time [e.g. a letter of 1687 in Dutens VI. 1, p. 36]. Bernoulli's originality is to see what the notion of certainty implies for probability. He thinks certainty is of two sorts, subjective and objective. Anything that will occur is already objectively certain. An historical dictionary gives a good reminder of what is going on here: the word 'certain' once meant what was decided by the gods. If some event were objectively uncertain then the gods could not have made up their minds. Perhaps in a more mechanistic age, if what happens does not happen with certainty then determinism is false. But Bernoulli is not much afflicted by causal determination of the sort made familiar from the time of Laplace. He explicitly says that causal, mechanical, determinism is not his problem. The 'omniscience and omnipotence of the Creator' guarantees the absolute objective certainty which derives from first causes. 'Some people argue about how the certainty of future existents can fit in with the dependence or independence of secondary causes, but since this issue has nothing to do with our goal we do not wish to touch on it.' The secondary causes are the efficient causes which are the only causes acknowledged by our modern conceptual scheme. He is worried only about final causes: 'divine supervision' and 'divine predetermination'.

Our individual judgements of certainty are in contrast to objective certainty. An event is certain relative to a given body of information if it is impossible both for the information to be correct and for the event to fail to happen. Probability is degree of this kind of certainty. Complete certainty of this sort can be obtained through demonstration of direct observation. Sometimes we can almost achieve complete subjective certainty. Then we have *moral certainty*. This concept was familiar even from casuistry. As early as 1668 Leibniz had connected it with probability, writing of what is 'infinitely probable or morally certain' [*S.S.* vi. 1, p. 494]. Bernoulli, however, says that 'infinitely little certainty' is the same as impossibility, whereas something with 1/1000 of certainty is morally impossible. We may consider something morally certain if it has 999/1000 parts of certainty. Just like modern statisticians who use 1% or 5% significance tests at will, Bernoulli is not dogmatic about the fraction 999/1000. What is important is rather a common standard. 'It would be useful if the magistrates set up fixed limits for moral certainty', whether it be 0.99 or 0.999, for 'then a judge could not be biased'.

Bernoulli undoubtedly imported the word 'subjective' into probability theory but by now the word in this connection has become equivocal. Several distinct modern theories of probability have been called subjective, and since there has been so much idle controversy about them it is worth getting the terminology straight. For my part I prefer to avoid the word 'subjective' altogether, so as to avoid potential ambiguity, but since Bernoulli started the trend we cannot evade it here. In recent writing three different kinds of probability have been called subjective. The most extreme subjectivism is that of Bruno de Finetti. L. J. Savage has felicitously called it 'personalism'. Then there are the theories of logical or inductive probability developed by J. M. Keynes and others, and which, though insisting on a measure of objectivity, are often called 'subjective' by their detractors. Thirdly there is yet another concept of subjectivity current among many present philosophers of quantum physics.

A convenient indicator is available to test an allegedly subjective theory. It devolves on the notion of unknown probabilities. In the extreme personalist theory, probabilities may be unknown only insofar as one 'fails to know one's own mind' (to use Savage's phrase.) In the logical theory probabilities may be unknown by failure to do logic but no experiment will help check up on logical probability. Finally in some physicists' use of the term 'subjective', subjective probabilities can actually be checked by experiment. We shall find that, contrary to the authorities mentioned in the beginning of this chapter, Bernoulli's subjectivism is less like the personalist or logical point of view, and more like that of the physicists.

To begin to use this let us pretend for a moment that in some set-up the chance of getting heads with a coin is an objective physical characteristic of the coin; pretend further (contrary to Bernoulli) that the chance is an ultimate fact, in that the outcomes of particular tosses are not determined by other features of the set-up and coin. Thus chances of heads are ultimate tendencies or propensities. Contemplating the outcome of the next toss, you say correctly that the probability of getting heads is p. Call this a 'pure case'. Now suppose that many coins, of different known chances of heads, are put in an urn. One is drawn at random. Assuming there is an equal probability of drawing each coin, we can compute a 'probability' of getting heads on the next toss of this promiscuously selected coin. Call this a 'mixed case'. The probability computed in the mixed case is not in general the tendency or the propensity of the coin to fall heads (even though it doubtless has to do with propensities of the tandem chance set-up consisting of urn and coin-tossing device).

Now contrast two statements of the form: 'the probability of getting heads with this coin is p'. One statement is made in the pure case, one in the mixed case. In the pure case, we are speaking of an objective tendency of the coin (or so I have pretended). But in the mixed case our probability statement, insofar as it is a statement about this coin, results in part from our ignorance about which coin it is. It corresponds, perhaps, to an objective statement about the probability of getting heads by first selecting a coin at random and then tossing it. But many writers have gone on to speak of the probability of getting a head from this coin, and this statement in the mixed case may naturally be called 'subjective'. Better, as Heisenberg puts it, 'the probability function contains the objective element of tendency and the subjective element of incomplete knowledge' [1959, p. 53]. Heisenberg writes not of my fiction of coins and urns, but of an essential feature of the quantum theory. As he interprets the theory, there exist systems in pure states whose probability function is 'completely objective'. But there also exist ensembles of systems in different pure states; these are called mixtures. Probability statements about a system in a mixture depend on 'statements about our knowledge of the system, which of course are subjective'. Subjective, yes, but no mere matter of opinion, for these so-called subjective probabilities can still, he says, 'be checked by repeating the experiment many times'.

De Finetti means something rather different. He views subjective probability as an indicator of purely personal degrees of belief and he has discovered useful constraints on them. For all his constraints I may in consistency have virtually any personal probability function for any given set of alternatives. Probabilities are unknown only insofar as I may fail to know my own mind.

Intermediate is the theory of logical or inductive probability according to which, any body of evidence e uniquely determines a probability for any hypothesis h. This is best represented as a function $c(h, e)$, meaning the degree to which e confirms h. Some early writers thought one could detach the evidence e: that is, if $c(h, e) = p$, and if we know just e, then we can say that the probability of h is p. This makes the probability of h a 'subjective' matter of what e we possess. But Keynes and others have insisted that logical probabilities cannot be detached from evidence. They maintain that probability is an objective logical relation between h and e. Hence they say this is no subjective probability at all. All the same, the label 'subjective' has not yet become completely unstuck. But the indicator of unknown probabilities is a pretty sure test of whether we have a logical probability. One may fail to know $c(h, e)$ only through failure to do probability logic.

It is certain that Bernoulli never entertained a fullblown personalist theory in the manner of De Finetti or Savage. According to him, the art of conjecturing aims at 'measuring the probabilities of things as exactly as possible'. He is concerned with what probabilities are appropriate to what evidence. This suggests P. M. Boudot's [1967] claim that Bernoulli held a theory of logical probability. That honour must be reserved for Leibniz, for although Bernoulli wavers somewhat, in his most important work at the end of Part IV Bernoulli is concerned with the experimental discovery of unknown probabilities. He thinks they are produced by some kind of symmetries and that the art of conjecture should produce good estimates of those unknowns. Heisenberg, in his physicist's conception of 'subjective probability', thinks that such probabilities can 'be checked by repeating the experiment many times'. Bernoulli was the first to investigate how many repetitions are required before we may be confident of our estimates.

There is no need to foist a single probability idea on to Bernoulli. Indeed it is in consequence of his work that the distinction between aleatory and epistemic concepts of probability became more important. He was interested in estimating unknown aleatory probabilities. He also wanted to know how certain he could be of any given statement about an aleatory probability. This invites the notion of epistemic probability. Thus we arrive at probabilities of probabilities – epistemic probabilities of statements about aleatory probabilities. However I find no evidence that Bernoulli made this step. As we shall see in the next chapter, some subsequent fallacious uses of Bernoulli's limit theorem do make the step, but there is no strong evidence that Bernoulli committed the fallacy and hence no strong evidence that he thought he had computed epistemic probabilities of aleatory probabilities.

In the early chapters of Part IV Bernoulli works chiefly in what we should now call the realm of epistemology. His basic problem is how to combine evidence of different sorts. In simplified models of games of chance there is a Fundamental Probability Set of equally probable cases. In real life we have a variety of evidence, some counting for a given opinion and some against it. Even today there is no very good way to combine different kinds of evidence into a single probability statement, and perhaps none will ever be discovered. Bernoulli had as good a try at this problem as anyone. It occurs chiefly in the second and third chapters of Part IV. It is rather closely patterned on the Port Royal *Logic*, and indeed uses examples such as the notaries, described in my Chapter 8 above. The discussion begins with Port Royal's crucial distinction between internal and external evidence. Bernoulli further distinguishes various kinds of proof. If we are curious to know whether Maevius killed Titius, we may (1) ask whether Maevius had a reason for killing Titius, (2) notice that Maevius shows the effects of such an action, e.g. turns pale on questioning, (3) look for signs of the event, such as a bloodstained sword, (4) discover if there is any circumstantial evidence, such as Maevius being on the same road when the killing occurred, and finally (5) take down any testimony.

The real problem of epistemic probability must be to combine types of evidence. Bernoulli did not succeed. First, he elaborates on Port Royal's rules for sound judgement, insisting that all evidence be accounted for and that we must sometimes suspend judgement. Unfortunately this sage counsel does not seem to take us any closer to combining these different kinds of evidence. That is left for the suggestive but ultimately abortive Chapter 3.

Bernoulli adopts a subtle and adaptable model. He explains it using a terminology of necessary and contingent events which is now rather foreign to us. A proposition is called necessary, relative to our knowledge, when its contrary is incompatible with what we know. It is contingent if it is not entailed by what we know. Suppose we are concerned with an argument, employing evidence e in support of hypothesis h. We disregard the necessary case, when e is known for sure and known to entail h. There remain three other cases. We may know that e entails h, but not be certain of e. This argument is contingent, and indicates h necessarily. Or we may be certain of e, which in turn only makes h probable. Then the argument is necessary and indicates h only contingently. Finally an argument may be contingent and indicate h only contingently. Modern inductive logic treats only the case of a necessary argument indicating h contingently, but in real life all three kinds of arguments are used. The modern attitude is a consequence of the foundationalist picture of knowledge; we are supposed to have some basic certain evidential knowledge e, from which we infer, perhaps only with probability, all the rest of our superstructure of beliefs. If we dissent from this foundationalist programme, we shall find that

The art of conjecturing

Bernoulli's classification provides a better model of much probabilistic reasoning.

There is a second scheme which cuts across the first one. Apparently it descends from Leibniz's notion of pure and mixed proofs in law, discussed in Chapter 10. Certainly the words are the same, although 'pure' and 'mixed' may be the inevitable words here. They are used, after all, by physicists describing quantum states, as quoted from Heisenberg earlier in the present chapter. I attribute Bernoulli's terminology to Leibniz because so much else in Bernoulli bears signs of the collaboration between the two men.

Mixed evidence is familiar. It breaks down into x cases that favour h, and y cases that count against h. Cases are assumed to be equally probable. Familiar games of chance employ mixed evidence, and so does much statistical data. Pure evidence breaks down into x cases that favour h and y that are simply neutral. Bernoulli's own examples are as follows. Someone is murdered in a mêlée. Some black-cloaked man in the crowd did him in. There were four men so clad; one is Gracchus. This is mixed evidence of which $\frac{1}{4}$ counts for his guilt, and $\frac{3}{4}$ against it. Suppose that on questioning Gracchus grows pale. This, says Bernoulli, is pure evidence, for if the pallor is sign of a guilty conscience it proves guilt, but other cases are possible. Gracchus may blanch out of embarrassment or grief. In such cases a whitened face is completely neutral as regards Gracchus' guilt.

Pure/mixed and necessary/contingent are two modes of classification which intersect. For example, suppose that in a transcontinental train I find a scrap of paper, datelined 'Monday . . .' but lacking the actual date, and predicting a blizzard 'with 80% probability' in the town of my destination on the day following printing. I find this fragment on a Monday night, and wonder whether I shall awake in a blizzard. I do not know whether this scrap is from this morning's paper, or from some other Monday. I am, let us say, $\frac{3}{4}$ certain that it is timely. That is pure evidence, because 75% of it bears on my conjecture, that I shall awake in a snowstorm. But in the other case, to which I attribute 25% of certainty, the newspaper is out of date and tells me nothing at all about tomorrow's weather at my destination. As well as being pure evidence, it indicates a blizzard only contingently: if this is today's paper, there is still only an 80% chance of a blizzard. It is tempting, and probably correct, to multiply the two probabilities to find the degree of certainty of a blizzard tomorrow: $\frac{3}{4}$ of 80% making me 60% confident. But what if I also have mixed evidence? Suppose my weather lore is sufficiently detailed to know that spring blizzards at my destination occur less than one day in fifty? This is mixed evidence which needs to be combined with pure evidence. It is not a problem discussed in current statistics or inductive logic, but it is, perhaps, one of the most common of everyday evidential situations. Bernoulli's own solution is rebutted by Lambert [1764, sec. 239] and the issues were further elaborated by Prevost and Lhuillier [1797]. I have studied Bernoulli's laws of combination in my [1974] and will not repeat the summaries here: suffice to say that Bernoulli's solutions are, although open to dispute, extremely instructive and perhaps right.

There is another aspect of his chapter that needs recording simply because in the annals of probability it has been entirely forgotten. This is the matter of additivity. Bernoulli sees clearly that probabilities derived from mixed evidence are additive. Frequencies are also additive: that is, the relative frequency of two mutually incompatible events, within some reference class, is the sum of their individual frequencies. There are also arguments for saying that degrees of belief, when represented by fractions, should be additive. Nearly all of these rely on construing a degree of belief as a betting rate, or as a more subtle measure of preference among actions. Bernoulli did not have this idea. As he said, his probabilities are degrees of certainty. Now consider pure evidence. Suppose, for simplicity, that I find a newspaper weather forecast, which I regard as infallible, and which categorically predicts a blizzard on the day after printing. I am only $\frac{3}{4}$ certain that it is today's newspaper. So I am only $\frac{3}{4}$ certain of a blizzard tomorrow. It does not follow that I am $\frac{1}{4}$ certain of no blizzard.

Our modern probabilist will say that $\frac{3}{4}$ marks a lower bound on the probability of a blizzard. My degree of belief, as represented, perhaps, by a betting quotient, should be $0.75 + p$, where p is some fraction less than $\frac{1}{4}$. My degree of belief in no blizzard is $0.25 - p$. That, however, is not the way Bernoulli thinks. If the weather forecast is all I have – I am otherwise a meteorological ignoramus in this part of the world – then the fragment of paper gives me $\frac{3}{4}$ of certainty and no more. It gives me no certainty whatsoever of no blizzard. Thus in Bernoulli's opinion, the probability of a blizzard, given this data, is 0.75, and the probability of no blizzard is 0.

Whether or not we call this a probability concept, or, to avoid confusion, give it another name, this conception seems to me to be perfectly coherent, although perhaps not so useful as an additive probability concept.

Bernoulli's probabilities are not degrees of belief but degrees of certainty. Thus even if events A and B are disjoint, the probability of A-or-B need not be the sums of the probabilities of A and of B. Even more surprising, Bernoulli admits that the probability of A and of its contradictory may both exceed $\frac{1}{2}$:

> If besides the arguments that count in favour of the thing, other pure arguments present themselves, which indicate the opposite of the thing, the arguments of both kinds must be weighed separately according to the preceding rules, in order that one may obtain a ratio between the probability of the thing and the probability of the opposite of the thing. Here it must be noted that if the arguments for each side are strong enough, it can happen that the absolute probability of each side notably exceeds half of certainty. Thus each of the alternatives is made probable, although speaking relatively one may be less than the other. So it can happen that a thing possesses $\frac{2}{3}$ of certainty and its opposite possesses $\frac{3}{4}$ of certainty. In this way, each of the alternatives will be probable, but nevertheless the thing is less probable than its opposite, in the ratio of $\frac{2}{3}$ to $\frac{3}{4}$, or 8 to 9.

Unfortunately this conception was not further explicated, and completely died out of probability theory.

17

THE FIRST LIMIT THEOREM

Chapter 5 of Part IV of *Ars conjectandi* proves the first limit theorem of probability theory. The intended interpretation of this result is still a matter of controversy, but there is no dispute about what Bernoulli actually proved. He takes for granted a chance set-up on which he can make repeated trials. There is a constant unknown chance p of 'success' S on any given trial. When n trials are made a proportion s_n of successes is observed. Bernoulli proves what is now called the weak law of large numbers: the probability of an n-fold sequence in which $|p - s_n| < \varepsilon$ increases to 1 as n grows without bound. Moreover, for any given error ε, he shows how to compute a number n such that the probability of getting s_n in the interval $[p - \varepsilon, p + \varepsilon]$, itself exceeds any given probability $1 - \delta$. In particular, if $(1 - \delta) = 0.999$, we have a moral certainty that s_n will fall in the assigned interval. For example if p is 3/5 then a moral certainty of error less than 1/50 is guaranteed by an n in excess of 25 550.

Bernoulli's proof is chiefly a consequence of his earlier investigation of combinatorics, for it proceeds by summing the middle terms in the binomial expansion. Notice that this result is a theorem of pure probability theory, and holds under any interpretation of the calculus. There is a familiar frequency interpretation of the weak law of large numbers. If the relative frequency of S is p, then the relative frequency of sequences whose error $|p - s_n|$ exceeds ε, is itself small, and decreases to 0 as n increases. There is equally a betting rate interpretation of the weak law. If one's betting rates are coherent, viz, satisfy the probability axioms, and if one is indifferent about the order of different trials, then if one bets at rate p on S at any one trial, one should bet very little on making n trials whose $|p - s_n|$ is in excess of ε: one's rate should tend to 0 as n increases.

Mathematics sufficient to guarantee such a result places Bernoulli in the pantheon of probabilists. Karl Pearson [1925] has criticized it on the ground that the limits Bernoulli discovered are rather gross. Better limit theorems including one discovered by Jacques' nephew Nicholas, are finer, but Bernoulli's work is of special importance because it is intended as the very foundation of the art of conjecturing. He shows that for given p the observed proportion of success, namely s_n, will tend to p as n increases. Suppose we do not know p. Can we still make some use of Bernoulli's theorem? The question is pressing because Bernoulli himself introduces his theorem precisely for those cases when we lack *a priori* knowledge of p.

The limit theorem comes in Chapter 5. Chapters 2 and 3 had dealt with problems of combining evidence, but still assume a given Fundamental Probability Set of equally probable cases. 'The number of cases is known in dice playing. There are manifestly as many cases as faces [. . .] because of the similarity of the faces and the laws of gravity acting on the dice, there is no reason why one face should be more likely than another, as might happen if the faces were of different shapes or made of some inhomogenous substance more heavy on one side than another.' [p. 223]. Likewise, in drawing lots from an urn filled with black and white slips of paper, all the lots are 'equally possible', because the numbers of papers of each kind are determinate and known, 'no reason can be perceived why this or that should come out more easily than another'. Unfortunately the problems of real life are less tractable. We have stable mortality statistics, but who can ever tell the numbers of diseases? Who can enumerate the parts of the body that are attacked by disease? Yet it is a plain fact that plague kills 'more easily' than dropsy, and dropsy kills 'more easily' than fever. We have statistical regularities but no Fundamental Probability Set. The same arises with irregular dice, and with those games whose outcome depends at least in part on the skill of the players. No F.P.S. can be perceived *a priori*.

We require a method of determining the 'number of cases' when mere reflection is not enough. 'What you cannot deduce *a priori* you can at least deduce *a posteriori*.' For example, if it is known that of 300 men who resemble men like Titius in age and constitution, $\frac{2}{3}$ died in a decade, then the 'number of cases' in which Titius must die in ten years is twice the number in which he survives. Such reasoning is familiar also in predicting the weather and in assessing the chances of two players in some match of skill. 'This empirical way of determining the numbers of cases by trials is neither new nor unusual, for the celebrated author of the *Ars cogitandi* [. . .] prescribes a similar method in [. . .] the last part of this work' [p. 225].

Despite the familiarity of this empirical method of reasoning, three tasks remain. First, it must be proven 'from first principles' that the method is valid. Second, we must discover whether there is an upper bound ($\frac{3}{4}$, say) to the certainty that can be achieved this way, or whether, as the number of trials increases, we approach a moral certainty. Third, we must investigate the actual number of trials required before we attain a given level of certainty. Bernoulli warns us that our estimates should not be 'accepted as precise and accurate'. They will be interval estimates, 'bounded by two limits'.

Bernoulli's exposition has a basic difficulty that has led to repeated misinterpretation. It is still a matter of controversy. In order to be clear about this matter we shall need to adopt a more modern jargon. Bernoulli plainly wants to estimate an unknown parameter p. His favourite example is the proportion of white pebbles in an urn. An *estimator* is a function F from data to possible parameter values, in this case, possible values of p. Bernoulli uses an *interval estimator* which maps given data onto a set of possible values of p, 'bounded by two limits'.

The outcome of n trials, s_n, is a random variable. For any estimator F, the estimate $F(s_n)$ is therefore also a random variable. When F is an interval estimator, $F(s_n)$ has as its values intervals which, it is hoped, will include the unknown chance p. If F is to be an informative estimator, we shall expect the intervals $F(s_n)$ to be rather narrow. Bernoulli considers estimators that estimate the unknown p to be in some narrow interval around the observed s_n.

When is F a good estimator? We cannot produce an informative estimator that is certainly correct. Short of such perfection, two very natural desiderata present themselves. It seems natural to demand that F should usually give the right answer. $F(s_n)$ is a random variable. Hence, far more often than not, n-fold trials on our chance set-up should give results s_n such that $F(s_n)$ includes the true, unknown, parameter.

It also seems natural to demand that for any particular result r, if we observe r and know nothing further about the set-up (and hence estimate p by $F(r)$) then, in the light of this result r we can be pretty sure that p is in $F(r)$.

These two desiderata are not necessarily equivalent. For

example, take the following estimator *G*: When the number of observed successes is not *n*/2, *G* is just like Bernoulli's estimator, and it estimates that *p* is around s_n. But when the observed proportion of successes is exactly $\frac{1}{2}$, *G* estimates that *p* is around 0. Now *G* satisfies our first desideratum, because *G* is usually right. But it does not satisfy the second desideratum, for there is one possible result (getting exactly 500 *S* in 1000 trials, say) for which I am confident that *G*(*r*) does not include *p*. Whenever I observe $s_n = \frac{1}{2}$, I know *p* is not in *G*(*r*).

It is useful to label our two desiderata. It seems reasonable to require that an estimator should be *usually right*. It also seems reasonable to require that our estimator should be *credible on each occasion of use*. These two desiderata are not identical, for *G* is usually right, but is not credible on each occasion of use.

I am insisting only on the trivial point, that the two desiderata are different. The happiest discovery would be that a 'most usually right' informative estimator was also an estimator that is 'most credible on each occasion of use'. Such a discovery of course demands a precise explication of both concepts, especially of the concept of credibility on each occasion of use. At present the prospect of such a discovery is bleak. The last analyses of what makes for credibility on each occasion of use yield desiderata that are incompatible with the desideratum of being usually right.

Bernoulli himself is ambiguous. He wants to know the number *n* such that we can be morally certain that our estimate of *p* (namely an interval around s_n) is very nearly right. Contrast two statements of moral certainty. (i) Being about to make *n* trials, I am morally certain that my estimator will give me the right answer. (ii) Having made *n* trials, and observed s_n, I am morally certain that my estimate *F*(s_n) is right. Now (i) is concerned with the *before-trial* virtues of an estimator, while (ii) is concerned with *after-trial* evaluation. For before-trial evaluation, the criterion of being usually right is probably enough. But for after-trial evaluation, we will need an estimator that is credible on each occasion of use. Thus if Bernoulli is concerned only with before-trial evaluation, and with making statement (i), then he may need only one of our desiderata. I think it likely that many of Bernoulli's readers took him to be concerned with after-trial evaluation, and that leads to fallacies.

So far I have tended to follow Bernoulli, being somewhat vague about different interpretations of probability. Inevitably, however, we come to consider his problem as one of estimating an unknown aleatory probability, or chance. Moreover, we wonder if he wanted to know the epistemic probability that a given estimate of chance was correct. We have slipped into calling this epistemic probability the 'credibility' or degree of believability. That is not Bernoulli's usage but it helps to make some distinctions. Suppose we want an estimator to be credible on each occasion of use. Then we require that for each possible outcome s_n, the credibility of the proposition, 'p is in $F(s_n)$', relative to the datum 's_n occurred', should be high. In shorthand, we require that for each s_n,

$$\text{Credibility } (p \text{ is in } F(s_n) \mathbin{/} s_n) > 1 - \delta \qquad (1)$$

for some small δ. Here the stroke (' / ') is the stroke of conditional probability, and may be read 'given that' or 'conditional on'. What is credibility? The most commonly proposed answer is that credibilities are probabilities. That is, they satisfy the probability axioms. Now under any interpretations of 'probability', Bernoulli discovered how to compute some conditional probabilities of the form

$$\text{Probability } (p \text{ is in } s_n \pm \varepsilon \mathbin{/} p). \qquad (2)$$

Perhaps because of the lack of any perspicuous notation for conditional probabilities, it may have been tempting to take the values computed for (2) as the values of,

$$\text{Probability } (p \text{ is in } s_n \pm \varepsilon \mathbin{/} s_n). \qquad (3)$$

Then one could use the values for (3) as a measure of the credibility, that p is in the interval $s_n \pm \varepsilon$, relative to the data, and so check that (1) obtains for Bernoulli's estimator.

Naturally (3) does not follow from (2). It is worth emphasizing that although (2) can be given any probability interpretation (frequency, or betting rate, or what you like) (3) is not open to a frequency interpretation when p is some definite, unknown, parameter, and s_n is some definite experimental outcome. Either p is in the interval, or it is not, and there is no 'frequency' with which this occurs.

If Bernoulli wanted to be sure that his estimator is good on each occasion of use, that is, *if* Bernoulli wanted to know about after-trial evaluation, and *if* Bernoulli thought that credibilities are probabilities, then he needed to compute values of expressions such as (3). He did not succeed. But there is not much reason to think that Bernoulli was directing his attention to after-trial evaluation. There is certainly very little evidence that Bernoulli ever wanted to infer (3) from (2). Indeed it is unclear that any of the great probabilists was tricked by notation into that fallacious inference.

Thomas Bayes' paper published half a century after the appearance of the *Ars conjectandi* is the first systematic attempt to compute values for (3). Richard Price, introducing Bayes' work to the Royal Society, alludes to the fact that although Bernoulli and De Moivre have attained important results, they have not shown how to compute (3). According to Price, a computation of (3) is essential to the solving of the problem of induction. Remember, however, that at the time Bernoulli wrote, the problem of induction had not yet been stated as a central problem of philosophy. According to Price, Bayes was trying to solve the problem of induction. One thing Bernoulli was *not* trying to do was to solve some publicized problem of induction, for when he wrote there was none. Bayes' solution to (3) has recently become very well known: it assigns prior probabilities over the parameter space and by means of 'Bayes' theorem' computes the posterior probabilities (3). If Bernoulli had wanted to be morally certain of his estimator on each occasion of use, then perhaps he should have invented Bayes' theory. That is, if he had wanted after-trial evaluation, perhaps he should have anticipated Bayes. But I repeat that there is little indication that Bernoulli wanted any such thing. For a fairly historical account of what Bayes did in his paper of 1763, consult my [1965, Ch. xii].

For any given value of p, Bernoulli is able to compute (2). Thinking of the special case in which $F(s_n)$ is the estimator $s_n \pm \varepsilon$, Bernoulli is thus able to compute, for each possible value of the unknown p, the probability of getting an s_n such that $F(s_n)$ includes that value of p. Let us call this probability 'the probability of being right'. On a frequency interpretation it is just the relative frequency of making an estimate that includes the true value. Bernoulli did not investigate the way in which the size of interval might be adjusted for different values of p. But his own mathematics is sufficient to observe the following. For a given number of trials n, and for given error ε, the probability of being right is a function of p, but considering all values of p in $[0,1]$, there is a minimum probability of

159

being right. Hence Bernoulli's results entitle him to make the following kind of statement:

Regardless of the true value of p, the probability that an (4) estimate $F(s_n)$ will include the true value, is at least $1 - \delta$.

Notice that this statement, although open to a degree of belief interpretation, is also open to a frequency interpretation. When ε is big enough for δ to be small, this statement entitles one to conclude that the estimator F is usually right.

Hence Bernoulli could show his estimator $s_n + \varepsilon$ is 'usually right', and is good for before-trial estimation. There are two elements to the solution. First, if I have a chance set-up, and I am about to make a trial of some kind, and the probability of getting result r on trials of that kind is at least $1 - \delta$, then I can (on Bernoulli's view) be morally certain that r will occur, so long as δ is small enough. Secondly, Bernoulli is entitled to assert (4); we take the special case, $r = {}'F(s_n)$ includes the true value'. Hence he can conclude, 'I am morally certain that my estimator will give me the right answer'. This is a before-trial evaluation of the estimator. Bernoulli is not entitled to conclude that his estimator will be credible on every occasion of use. His mathematics is not adequate for after-trial evaluation of particular estimates. He can say, 'I am morally certain that this estimator will give me the right answer.' He can make an experiment and say 'My estimator applied to this observation gives me this estimate.' But he cannot validly conclude, 'I am morally certain that this estimate, got by applying my estimator to this data, is a correct estimate.' As observed earlier, credibility on an occasion of use does not follow from an estimator's being usually right. It may, perhaps, be just a lucky choice of words but Bernoulli does on occasion give the impression of having only before-trial evaluation of estimators in mind. Speaking of his favoured example, drawing from an urn of whose pebbles are white, he told Leibniz that 'you can be morally certain that the ratio obtained by experiment will come as close as you please to the true ratio of 2: 1.' [Leibniz *M.S.* III. 1, p. 88]. This appears to be before-trial statement. He does not claim that after the experiment we can be morally certain that our estimate is as close as we have designated to a true unknown value.

A statement like (4) will show one that an estimate is good from the point of view of being usually right, but it does not show that it is best, from that point of view. In fact (4) is in an obvious way inefficient. Let us call $1 - \delta$ the *security level* of our estimator $F(s_n)$. It would be more efficient – more informative – if we obtained the estimator with the smallest intervals compatible with a given security level. This requires an investigation of how size of interval is a function of different values of p. It is not a difficult problem to conceive. Indeed O. B. Sheynin [1968] has pointed out that Nicholas Bernoulli, Jacques' nephew and posthumous editor, worked out a special case. The solution is, however, tucked away in correspondence with Montmort, in a letter of 23 January 1713, and published in the second edition of Montmort's book on games of chance. A more general treatment requires much more than the binomial computations of the Bernoullis and more even than De Moivre's normal approximation of the binomial distribution. E. L. Lehmann [1958] suggests that Laplace was the first to produce a general solution to what I have been calling Bernoulli's problem. On p. 287 of the *Théorie Analytique* Laplace appears to compute estimators F_δ with the property that

For all p in $[0,1]$: Probability$\{p$ is in $F_\delta(s_n)\} = 1 - \delta$. (5)

This is the universal quantification of an expression like (2). It may look as if p is a random variable, but it is not. In (5) we say that regardless of the true value of p, the probability of making a right estimate is $1 - \delta$. This $1 - \delta$ in (5) is thus the exact security level. Unfortunately, as Glen Shafer has pointed out to me in a letter, Laplace does not really obtain (5) because at one point he has to substitute the observed s_n for the unknown p, and hence the solution is only asymptotically correct. W. S. Gosset's famous statistic 't' was perhaps the first device to overcome this kind of inexactness. It was not available until 1908. However, if we ignore this kind of inexactness we can regard the theory of 'probable errors' produced by Gauss in 1816 as using interval estimators with a security level of 0.5.

A security level of $1 - \delta$ does not imply that F_δ is credible to degree $1 - \delta$ on each occasion of use. Elsewhere Laplace did investigate the problem of credibility of estimates made after observations, and presented a straightforward Bayesian analysis. [*Oeuvres*, VII, p. 371]. His two treatments are typographically separate, having a hundred pages between them, but one regrets that he did not enunciate the conceptual distinction between them.

Had he been more attentive to philosophical niceties he might have been spared some criticism. The astute logician De Morgan [1838], for example, does not seem to notice the difference between (2) and its universal quantification (5). He accuses Laplace of fallaciously inverting (2) to obtain (3). Reading De Morgan, it looks as if Laplace suffered from inadequate notation for conditional probability. Misunderstanding of Laplace more likely results from lack of quantifiers. Laplace is not mistakenly inferring (3) from (2) by not noticing what he conditions on. He is only stating (5), and hence asserting a security level for his estimator.

Although Laplace found an estimator which at least asymptotically has an exact security level, it is not unique. Hence other desiderata are required to choose among the estimators of given security level. The best known solution is due to Jerzy Neyman. When the security level is 0.999, the estimator gives an interval including the true value of p 99.9% of the time. But what if it includes wrong values of p even more frequently? That seems undesirable. So one would like to minimize the chance of including false values of p while maintaining a given security level. The Neyman theory results from applying this desideratum. In many interesting situations there exists a unique estimator of given security level that for every false value of p minimizes the chance of including p in the interval estimate. An interval estimate got from such an estimator is a confidence interval. As Neyman has always insisted, if I is a particular 99% confidence interval got from some observations, one cannot say, 'it is 99% probable that I includes the true value of the parameter'. One cannot automatically assess the credibility of any particular estimate, for one can only assess the long-run reliability of a system of estimation. According to Neyman, inductive inference is impossible. We must, he says, be content with inductive behaviour. We can behave in a way that is usually right, but we cannot measure the credibility of our doing the right thing on any individual occasion. This is one of the chief bones of contention in contemporary philosophy of statistics. The Bayesian school, for example, has quite the opposite opinion.

Only in the 1930s did Neyman and E. S. Pearson make clear the logic of the confidence interval approach. It would be foolish to contend that Bernoulli grasped the principles that have only been enunciated recently. We are, however, confident that Bernoulli did not make any simply fallacious 'inverse' use of his theorem, inferring (2) from (3). Taking the example of an urn which, unknown to us, has 3000 white balls and 2000 black ones,

it will be shown that for any given probability [i.e. any $1 - \delta$], it is more likely that the ratio obtained by frequently repeated trials with replacement will fall within an interval around 3 : 2, rather than outside that interval.

And that is just what Bernoulli proved. He thought it had application to inverse inference, but does not make clear exactly why. I have outlined some later developments as one possible consistent development of his line of thought.

It would have been nice to quiz Bernoulli from our present standpoint. We have to content ourselves with Leibniz's queries presented in letters at the end of 1703. Leibniz, alas, could not ask our questions. Nevertheless, he is shrewd, and Bernoulli stands up well to this grilling. Writing to the philosopher on 3 October 1703 Bernoulli made perhaps the strongest claim for his theorem:

I can already determine how many observations must be made in order that it is 100 times, 1000 times, 10 000 times, etc. more likely than not – and this is moral certainty – that the ratio between the number of cases which I estimate is legitimate and genuine [i.e. within some allowed error. Leibniz, *M.S.* III. 1, p. 78].

This sounds as if Bernoulli has an inverse application of his theorem in mind, but writing on 20 April 1704 he gives a more accurate statement of what he could do. Leibniz is worried by Bernoulli's claims but his are not the problems of the modern statistician. He is troubled by the lack of an available Fundamental Probability Set in the cases of *a posteriori* estimation. He grants that there is an *F.P.S.* for dicing and urns, but cannot persuade himself that there is one for diseases or 'changes of the air'. Bernoulli made two replies, one in a letter, and one in the *Ars conjectandi* itself. The difference between the two answers is instructive. In the book he gives an epistemological answer. Comparing urns and diseases he says that 'with respect to our knowledge both numbers – the numbers of diseases and the number of pebbles in an urn – are equally uncertain and undetermined' [p. 204]. However, in a letter of 1704 he replies in a much more aleatory vein:

If now in place of the urn you substitute the human body young or old, which contains the tinder of diseases like the urn contains stones, you can in the same way determine how much nearer the old man is to death than the young one.

Leibniz also objects that whereas the number of stones or whatever is finite, the number of diseases is unbounded. Bernoulli replies with the notion of limit, explicitly referring to the newly discovered fact that π is the limit of a sequence of ratios. No wonder that von Mises, with his analysis in terms of probability as the limit of a proportion within a particular kind of sequence, should take Jacques Bernoulli as one of his own!

Of course no one is less likely than Leibniz to be scared of infinity. His deeper worry arises from his whole conception of *a posteriori* learning. He observes that given any finite number of observations of, say, the path of a comet, infinitely many curves can be found to fit. Equally in the case of a finite number of trials: any number of statistical hypotheses will conform to the facts. The point is not that any number of statistical hypotheses are logically consistent with a finite experimental segment. The point is far stronger. Bernoulli shows that when the observed proportion of heads in n tosses is s_n, there would be a very good probability of getting s_n if the true unknown probability of heads were itself close to s_n. Apparently this is a reason for estimating the unknown p as around s_n. But there are infinitely many arbitrary hypotheses on which s_n would be just as probable. This is a traditional difficulty that is still regularly aired. Bernoulli's response is crisp. When in doubt, choose the simplest hypothesis.

Finally Leibniz protests that the number of diseases can hardly be supposed constant over the course of time. 'It is certain that someone who tried to use modern observations from London and Paris, to judge mortality rates of the Fathers before the flood, would enormously deviate from the truth.' Bernoulli replies with aplomb that nothing follows from this, except that occasionally new observations are to be adopted, 'just as they would be adopted if the number of stones in the urn were supposed to change'. It is a curious footnote that Leibniz's great contemporary, Newton in the posthumously published [1728], had actually tabulated biblical chronology and compared it with more recent history, and obtained an estimate of the duration of the reigns of kings that fitted both England since the conquest and Israel before the captivity. Karl Pearson's [1928] shows how excellent was Newton's analysis. I have said nothing of Newton in this book because probability so seldom engaged his attention. O. B. Sheynin's [1971a] collects a number of his occasional remarks on the subject. But although Newton's direct contribution to the understanding of probability was insignificant, his indirect influence may have been great. This is the topic of our next chapter.

18

DESIGN

The story of the emergence of probability comes to an end with the publication of *Ars conjectandi*. In 1711, even before the book appeared in print, Abraham de Moivre published *De mensura sortis*, which soon was to culminate in *The Doctrine of Chances*, where the mathematics of probability was recognized as an independent discipline in its own right. We have only one task left: to describe certain philosophical positions that are consequent upon the events described in preceding chapters. One of these is the sceptical problem of induction, published by Hume in 1739, and the other is the problem of chance in a deterministic universe. Although the former is epistemological and the latter arises from aleatory concerns, they are no more independent than any other bifurcation in the dual concept of probability. Determinism is, however, the one to study first.

The most immediate significance of Bernoulli's limit theorem lies not in a distant potential for sound statistical inference but in making more intelligible the sheer fact of statistical stability. A curious 'pre-Bernoullian' paper of 1710, by John Arbuthnot, usefully illustrates this fact. Arbuthnot is now chiefly known as a satirist esteemed by his contemporaries next only to Jonathan Swift. He was also Queen Anne's doctor, a Fellow of the Royal Society, and an amateur of mathematics. In 1692 he published the first English translation of Huygens' textbook. As Todhunter says, 'the work is preceded by a Preface written with vigour but not free from coarseness' [1865, p. 50]. The examples are characteristic of a bawdy age. To make the point that most of our assessments of probability are *a posteriori*, Arbuthnot reflects on the odds that 'a woman of twenty has her maidenhead', or that 'a town-spark of that age "has not been clap'd"'. He displays a sensible use of published statistics: 'it is odds, if a woman be with child, but it shall be a boy, and if you would know the just odds, you must consider the proportion in the Bills that the males bear to females'. He made no theoretical speculation as to why the observed proportion of boys should always exceed that of girls. Apparently the question lay fallow until almost two decades later when he published 'An argument for divine providence, taken from the constant regularity observed in the births of both sexes'. Although published in the *Philosophical Transactions* for late 1710, the paper must have been printed in 1711, for it includes data on births going to the end of the preceding year.

The 'Argument' employs two pieces of statistical reasoning, only one of which is valid, and it also claims to infer the action of divine providence from statistics. So we must examine three inferences, only two of which would now be called statistical, and the third of which is about the very nature of statistical stability. Arbuthnot considers a population of n coins whose sides are marked M and F. The binomial coefficients in the expansion $(M + F)^n$ give the probabilities of outcomes of the n tosses. Thus the coefficient of $M^k F^{n-k}$, divided by 2^n, is the probability of getting exactly k M and $(n-k)$ F. As n gets large the coefficients of any term get small 'and consequently (supposing M to denote male and F female) that in the vast number of mortals, there would be but a small part of all the possible chances, for its happening at any assignable time, that an equal number of males and females should be born.'

So far so good. In particular if a man wagers in tossing a large even number of coins that he will get exactly as many M as F, his chances are vanishingly small. But then Arbuthnot contends that the same holds even if he is wagering to get *approximately* the same number of M and F. 'It is very improbable (if mere chance governed) they [the outcomes] would never reach as far as the extremities.' That is, it is very improbable that we would not sometimes get a vast preponderance of M over F, or F over M. But in fact the Bills of Mortality show that no such thing ever occurs. Therefore the 'constant regularity in the births of both sexes' cannot be a matter of chance.

This argument is invalid. It is true that it is 'very improbable' that the outcomes 'would never reach as far as the extremities'. But unlike Bernoulli, Arbuthnot was unable to quantify the qualitative 'very improbable'. If he had, he would have found, as Nicholas Bernoulli subsequently showed, that the constant regularity observed is exactly what one would expect if the chance of a male birth is 18/35. Although there is always a positive probability of getting an extreme outcome it rapidly diminishes as *n* gets large. Even in 1692 Arbuthnot had noticed the slight surplus of males over females, but instead of reasoning like Bernoulli he tries to put this fact to his own purposes:

To judge of the wisdom of the contrivance, we must observe that the external accidents to which males are subject (who must seek their food with danger) do make a great havoc of them, and that this lot exceeds that of the other sex, occasioned by diseases incident to it, as experience convinces us. To repair that loss, provident nature, by the disposal of its wise creator, brings forth more males than females, and that in almost constant proportion.

The thought leads Arbuthnot to a second calculation. If the chance of a male birth were exactly 1/2, what would be the probability that, in any year, more males should be born than females? Almost exactly 1/2. Call this the probability of a 'male year'. The bills of mortality record 82 successive male years and no female years. The probability of this event, on the hypothesis of equal chance, is $(1/2)^{82}$, namely odds of 1 to 483 600 000 000 000 000 000 000. Moreover, if we consider that the surplus of males lies within quite narrow limits and moreover, as far as we know, does so 'for ages of ages, and not only at London, but all over the world', the probability approaches an infinitely small quantity. Hence the hypothesis of equal chance must be rejected. This has been called the first published test of significance of a statistical hypothesis.

In broad outlines, this argument is valid. This is in contrast to the first argument, which was based on an inadequate understanding of the limiting properties of chances. Then Arbuthnot has a third, 'metastatistical' argument. Since the constant proportion cannot be due to equal chance, in the matter of birth it must be 'art, not chance, that governs'. Nicholas Bernoulli gave the correct retort:

Let 14 000 dice, each having 35 faces, 18 white and 17 black be thrown up, and the odds are very great indeed that the numbers of black and white faces shall come as near, or nearer to each other, as the number of boys and girls in the bills [Montmort, 1713, a summary of pp. 388–90].

There may be two elements in Arbuthnot's misapprehension. There is his first, invalid, argument. He is simply ignorant of the fact that an event of chance *p*, on sufficiently many repeated trials, will very probably occur with a relative frequency very close to p. It requires no 'art' to guarantee this consequence of Jacques Bernoulli's limit theorem. A second factor may be the reliance, in the early days, on a discernible Fundamental Probability Set of equally probable cases. It might take experiment, rather than mere reflection, to establish that in throwing dice it is permutations, not partitions, that are equally probable, but once we do have some experience, we can actually 'see' the *F.P.S.* of cases. If, as was thought, experience teaches that a uniform mortality curve is a good fit, we can actually discern the equally probable cases, namely years. But in the matter of births, there is no discernible *F.P.S.* of $35k$ elements, $18k$ of which favour *M*. For lack of an *F.P.S.*, there can be no 'chance'. This adds further credence to the view that art, not chance, governs.

Arbuthnot's slender paper had a remarkable success. It was a tiny but influential contribution to the work of a group of men who endeavoured to relate Newtonian science to natural religion. Anders Jeffner aptly calls them 'Royal Society theologians' [1966]. They were Fellows of the Royal Society, dabbled in mathematics, and recorded all kinds of minor observations of nature and experiment. They were also divines, usually complacent, comfortable and established, indifferent or even hostile to the evidence of revelation, and convinced that the new science is itself a witness to the Deity's handiwork and therefore to his existence. Their first representative is John Wilkins, first secretary to the Royal Society, whose thoughts on natural religion and probability are described at the end of Chapter 9 above. The Society bred more of the same, and they came to dominate the intellectual life of Britain in the early eighteenth century. Their most notable pulpit was furnished by the Boyle lectures, endowed by the will of Robert Boyle to provide 'proofs of the Christian Religion against atheists' and other notorious infidels.

We shortly examine the overall significance of this trend for the comprehension of probability, but first a specific interaction with Arbuthnot. His paper was printed in 1711; later in the same year William Derham commenced the third series of Boyle Lectures, later published as *Physico-Theology*. The title is as exact as that of its sequel, *Astro-Theology*. The lectures argue that the world is so manifestly well arranged in every particular that, like an intricate piece of clockwork, it demands an artisan. Derham was especially well qualified to recognize clockwork: indeed his first published book was a survey of different methods of clock-making [1696]. In the preface to *Physico-Theology* he states his aim: 'Having the honour to be a member of the Royal Society, as well as a divine, I was minded to try what I could do towards the improvement of philosophical matters to theological uses.' Among the evidences of design he notes that the

surplusage of males is very useful for the supplies of war, the seas and other expenses of the men above the women. That this is the work of Divine Providence and not a matter of chance, is well made out by the very laws of chance by a person able to do it, the ingenious and learned Dr Arbuthnot [1713, p. 176, n. 8].

Derham does not repeat Arbuthnot's arguments, perhaps because he is not 'a person able to do it', but augments the reasoning with some rhetoric. 'What can the maintaining throughout all ages and places these proportions of mankind, and all other creatures, this harmony in the generations of men, be but the work of the one that ruleth the world?' Or again, 'How is it possible by the bare rules and blind acts of nature, that there should be a tolerable proportion, for instance, between males and females?' [*Ibid*, p. 178].

It is important to distinguish three distinct intertwined questions. First, there is the question of whether a constant statistical stability can be the effect of chance. Following Arbuthnot, Derham undoubtedly thought not. That was a mistake, and Nicholas Bernoulli, among others, said so at once. Second, there is the question of why the chance of a male birth should be about 14:13 (as Derham calculated) or 18:17 (as Bernoulli observed). Why a slight surplus of males over females? Since the gathering of food, the manufacture of implements, the travels of commerce and the perils of war do more to harm the young male population, than the diseases peculiar to young women hurt the female population, it is good to have more males than females in order that every person shall have a mate. Hence the fraction 18:17 is itself evidence of divine providence. In the final edition of *The Doctrine of Chances* De Moivre had the last word. He thought that although Arbuthnot was a little obscure, he little deserves the strictures of Nicholas Bernoulli, for he,

might have said, and we do still insist, that 'as from the observations, we can, with Mr Bernoulli, infer the facilities of production of the two sexes to be nearly in a ratio of equality; so from this ratio once discovered, and *manifestly serving to a wise purpose*, we conclude the ratio itself, or if you

Design

will the *form of the die*, to be an effect of *intelligence* and *design*.' As if we were shewn a number of dice, each with 18 white and 17 black faces, which is Mr Bernoulli's supposition, we should not doubt that those dice had been made by some artist; and that their form was not owing to *chance*, but was adapted to the particular purpose he had in view [1756, p. 253].

In addition to the invalid argument for design, based on ignorance of Bernoulli's theorem, and the viable argument, based on the existence of a constant chance of 18/35, there is a third and more elusive consideration that underlies Arbuthnot's paper. It has to do with the very nature of chance. On the continent, and especially in France, the investigation of chance phenomena was given a notable tincture of subjectivism. Everyone agrees that there is no such thing as real chance but this fact can be explained in different ways. On the continent, to talk of chance is to talk of lack of knowledge. In England, chance is lack of skill. As Arbuthnot put it, 'It is impossible for a die, with such determined force and direction, not to fall on such a determined side, and therefore I call that chance which is nothing but want of art.' This gives some insight into the idea that 'art not chance' governs in the matter of births. They are skilfully arranged to appear in constant proportion, and so a matter of art. Arbuthnot's opinion was partly based on ignorance of Bernoulli's theorem, but even De Moivre, better placed than anyone else to know the power of laws of large numbers, ventured into this domain. He had no use for the various words in frequent use, such as '*fate, necessity, nature, a course of nature* in contradistinction to the *divine energy*' [1756, p. 253]. If we attend to the phenomena and 'if we blind not ourselves with metaphysical dust, we shall be led, by a short and obvious way, to the acknowledgement of the great MAKER and GOVERNOR of all; himself *all-wise, all-powerful* and *good*'. It has been conjectured that De Moivre, despite his limit theorem, thought that statistical regularity still required a Divine hand to work. In a famous and often quoted passage Karl Pearson says that,

Newton's idea of an omnipresent deity, who maintains mean statistical values, formed the foundation of statistical development through Derham, Süsmilch, Niewentyt, Price to Quetlet and Florence Nightingale [. . .] De Moivre expanded the Newtonian theology and directed statistics into the new channel down which it flowed for nearly a century. The causes which led De Moivre to his 'Approximatio' or Bayes to his theorem were more theological and sociological than purely mathematical, and until one recognizes that the post-Newtonian English mathematicians were more

171

influenced by Newton's theology than by his mathematics, the history of science in the 18th century – in particular that of the scientists who were members of the Royal Society – must remain obscure [1926].

It is a 'statistical law' that the chance of a male birth is about 18/35. In consequence a large population will virtually always exhibit a slight surplus of male births. A statistical law is, we might say, about 'the course of nature'. It is just this that De Moivre claims is a vain and empty word. The proper concept is divine energy. In order to understand such unfamiliar thoughts, we must extend our domain, and look briefly at the very concept of law of nature in post-Newtonian Britain. The impact of Newton on this period was twofold. One was theological. He invested more labour on religious questions than on physics. Although his reflections were unpublished they were not unknown, and we find important traces of them, for example, in correspondence with Richard Bentley, the first of the Boyle lecturers. The theology was significant, however, chiefly because it came from the great Newton, whose conception of gravitational force was itself taken to have theological implications. There are also a number of uses to which Newton put probabilistic ideas but they are not our concern here. They are ably described in Sheynin [1971a]. Here it is his overall metaphysics, not Newtonian work on chances, that we must study.

Gravitation was the greatest discovery and the greatest mystery of the age. The initial successes of seventeenth century science had relied on 'mechanical' explanation. Gravitational force changed all that. The exact meaning of 'mechanical' is obscure. In his preface to Newton's *Principia*, Roger Cotes [1713] wrote that 'a quality is said to be mechanically caused when it is produced by some of the other affections of body'. George Cheyne, one of the Royal Society theologians (and one whom De Moivre accused of filching his ideas) said that something is mechanical when it can be explained by the three laws of motion. 'It is plain', he said, 'that these three laws do virtually comprehend all the rules of mechanism, and consequently, if any contradict these laws, or their necessary consequences, it is not to be mechanically accounted for.' Leibniz had the courage of this conviction and rejected the gravitational theory just because it abandoned the most successful research programme of the time, namely the investigation of nature by mechanics. To Samuel Clarke, the very dean of Royal Society theologians, he sneered at the new 'attractions' that 'some have begun to revive under the specious name of forces, although they bring us back into the Kingdom of Darkness' Clarke, in reply, said that by attraction we mean 'barely the effect or the phenomenon itself [...] whatever be the cause of it'. He was however prepared to abandon mechanism – whatever that is: 'the means by which two bodies attract each other may be invisible and intangible, and of different nature from mechanism, and yet acting regularly and constantly may well be called natural'.

Newton himself was cagey. In 1693 he told Bentley that 'gravity must be caused by an agent constantly acting according to certain laws, but whether the agent be material or immaterial, is a question I have left to the consideration of my readers' [Bentley 1842, p. 70]. He did think that some aspects of the solar system invited speculation about God. 'I do not know of any power of nature', he wrote to Bentley, 'that could cause the transverse motion of the planets without this divine arm.' The thought was often repeated. After the publication of *Astro-Theology*, Derham claims that Newton wrote him of a 'peculiar sort of proof of God':

He said there were three things in the motions of the heavenly bodies, that were plain evidences of omnipotence and wise counsel. 1. That the motion impressed upon these globes was lateral, or in a direction perpendicular to their radii, and above them in parallel with them 2. That the motions of them tend the same way. 3. That their orbits all have the same inclination. [Quoted in Manuel 1968, p. 127.]

Specific statistical investigation of the three points is of some importance throughout the eighteenth century. Starting with Daniel Bernoulli around 1730, the culmination is Laplace's work on the probability of causes, which began as an integral part of his celestial mechanics (as the Newtonian discipline had come to be called!). Such work, together with the theory of errors, is undoubtedly the staple of any history of probability for the period. Here we are concerned with the subtler but more fundamental philosophical vibrations of the Newtonian outlook. Whatever Newton thought about gravity, the Royal Society theologians made much of the notion that the laws of gravity are merely devices for computation, prediction, and description of constant regularities. They do not state the efficient causes by which bodies attract each other; they are merely constant conjunctions based on experience. It is at least plausible that we cannot know what constitutes gravitational attraction; it is at least possible that the actual efficient cause, whose effects we know through the constant and regular succession of events, is God himself. God did not merely wind up the clock of the universe, He is also constantly pushing all the objects together, according to certain fixed laws. By the time of Laplace all such talk seemed irrelevant. Efficient causes are nothing but laws of nature. But for De Moivre, the efficient cause was the regular action of 'divine energy'. This applied equally to chance and gravity. We have statistical laws, which, by limit theorems entail that in all probability constant proportions will be observed. But to say that the chance of a male birth is 18/35 is to describe the facts of nature in a compact way; it is not to explain the individual births that reveal this regular chance. Like gravity, the laws of chance are 'merely descriptive'. The causes lie in divine energy. This thought prompted Pearson's famous but cryptic utterance on the effect of Newtonian theology on chance.

Even when the importance of theological questions waned, this post-Newtonian attitude exercised a powerful effect. It opposed a subjectivist understanding of chance. The laws of chance state the real course of nature. Yet they are merely descriptive, and are in a sense gross laws, which ignore the fine structure of nature. We require a causal mechanism that could generate statistical laws. We understand that a chance of say 18/35 entails certain statistical stability, but we do not understand what, in the fine structure of the universe, creates the statistical stability summed up by the number '18/35'. One of the most important attempts to clear up these matters came to fruition with A. A. Cournot in the mid-nineteenth century. Cournot thought we could account for chance by independent causal chains. The most famous results in this tradition are by Henri Poincaré. There is also a less well known tradition of showing how Gaussian curves, either of population or experimental error, result 'inevitably' from some simple postulates about large collections of small unrelated events. Only the advent of the quantum theory has made it possible to conceive of statistical regularity as a brute and irreducible fact of nature. And there is still a small but valiant programme of 'hidden variable' theorists who contend that the gross statistical laws must be the manifestion of some as yet unspecified deterministic laws.

Thus the belief in an omnipresent deity that maintains mean statistical values has a strong and lasting effect on the aleatory side of probability. The central doctrine is that statistical laws merely describe constant regularities. Just like gravity, they do not get at efficient causes. This conception of 'mere regularity' is important not only for the aleatory side of probability, but also for the epistemic side. It is a final ingredient for the sceptical problem of induction, stated by Hume in 1739.

19

INDUCTION

The sceptical problem about the future, often called the problem of induction, was first published in 1739, in David Hume's *A Treatise of Human Nature*. It doubts that any known facts about past objects or events give any reason for beliefs about future objects or events. A similar problem arises also for inference about unremembered past events, and unobserved present ones, but I shall adopt Hume's own format. Will this bread nourish me? Hume argues that no collection of past observations on alimentation give any reason at all for thinking that the next piece of bread will also prove nourishing. Our expectations are formed by custom and habit, but lack justification.

Closely related is the sceptical problem about generalizations. Can any number of observed instances, short of a complete survey, ever make it reasonable to believe a generalization? The work of Hume has itself lent some credence to the view that particular predictions must be based on sound generalizations. Many philosophers think this problem equivalent to the problem about the future. Whether or not we agree with this supposedly Humeian doctrine, when it is not necessary to distinguish the two problems, we may speak simply of the sceptical problem about induction.

The *sceptical* problem is not to be confused with what may be called the *analytic* problem. Clearly people do distinguish good inductive reasons from bad ones, so we may begin to classify the various degrees of evidential support. This analytical task has been very substantially advanced in the twentieth century by philosophically minded statisticians. Clearly their predecessors broached the same problem long before Hume. Bernoulli did so in the fourth book of *Ars conjectandi*. Leibniz had a vision of inductive logic. Arguably Pascal also wanted to analyse non-deductive inference. We have quoted Hobbes, as early as 1640, 'if the signs hit twenty times for one missing, a man may lay a wager of twenty to one of the event'. That is at least a preliminary quantitative analysis. All these writers took for granted that, in Hobbes' words, 'they shall *conjecture best*, that have *most experience*, because they have most signs to conjecture by'. Slowly it was recognized that mere quantity of experience is not enough. The careful design of experiments can yield more food for conjecture in a week that the passing show of signs can deliver in a lifetime. But there is no doubt in anyone's mind that some signs do give good reason for beliefs about the future and about the unwitnessed past. Hume's sceptical doubts were unknown before 1739. Why?

The question is particularly pressing for the historian of probability because of what appears to be Hume's own view of the matter. In 1740 he published an anonymous advertisement for the *Treatise*. In this *Abstract* he tells us that,

The celebrated Monsieur Leibniz has observed it to be a defect in the common systems of logic that they are very copious when they explain the operations of the understanding in the forming of demonstrations, but are too concise when they treat of probabilities, and those other measures of evidence on which life and action entirely depend and which are our guide even in most of our philosophical speculations [   ] the author of *A Treatise of Human Nature* [...] philosophers [Arnauld, Malebranche, Locke] and has endeavoured, as much as he can, to supply it [1740, p. 7].

It may be that Hume was merely appealing to the current vogue for probability. The arguments from design, which originated with John Wilkins in the birthtime of probability, had culminated in Joseph Butler's *Analogy of Religion* in 1736, although the defective logical form of these arguments, cast as the character Cleanthes in Hume's *Dialogues Concerning Natural Religion*, was yet to appear. Butler, with a vastly greater audience than Hume, had already told the world it was not his 'design to inquire further into the nature, the foundation and the measure of probability [. . .] This belongs to the subject of logic, and is a part of the subject which has not yet been thoroughly considered' [1736, p. iv]. With less circumspection than Butler the same divines, who debated what proportion of revelation and what proportion of natural argument should be allowed in the foundation of religion, were incessantly quarrelling over how much probability to attach to the testimony of miracles in various epochs. Hume had already written his essay *On Miracles*, but kept this bombshell secret until [1748], when it prompted more critical study in the next two years than his work on induction was to receive for a century. So perhaps in his *Abstract* Hume was giving vent to justifiable pride that he understood the probability of design and of testimony far better than any contemporary. Or perhaps he was merely pandering to the current penchant for probabilizing. But I think Hume also thought that he could present his problem of induction by grace of his thorough grasp of probability. If so, why should not Hobbes, in 1640, have thought a little harder and propounded the same problem? Are we to suppose that what is commonly acknowledged as one of the great landmarks of epistemology occurred almost at random, and could as well have happened any time in the preceding century?

Is it not entirely clear, however, that the sceptical problem is Hume's. If we are liberal in our interpretations, we can, of course, always find anticipations and precursors. The most likely is the brief discussion in Sextus Empiricus' second century *Outlines of Pyrrhonism*. I quote Book ii, chapter xv in full:

It is also easy, I consider, to set aside the method of induction. For, when they propose to establish the universal from the particulars by means of induction, they will effect this by a review of all or some of the particular instances. But if they review some, the induction will be insecure, since some of the particulars omitted in the induction may contravene the universal; while if they are to review all, they will be getting at the impossible, since the particulars are infinite and indefinite. Thus on both grounds, as I think, the consequence is that induction is invalidated.

This might be read out of context as the sceptical problem about generalizations, but in fact it occurs in the course of a long discussion of demonstrative proof. Sextus has (like J. S. Mill's *System of Logic* one and a half millenia later) been accusing the syllogism of committing a *petitio principii*. If we use a premise, 'All *A* are *B*', to prove that this *A* is *B*, we must be begging the question. To ward off the objection that one might obtain 'All *A* are *B*', in some other way, he points out that induction is invalidated. There is no demonstrative proof, but Sextus does not, in this passage, contend that there is no reason, nor that inductive reasons are not reasons. Indeed he is seemingly content with much inductive inference. He strongly opposes those who favour the indicative sign, by which we infer something that is in principle unobservable from what has been observed. He had no truck with theoretical entities. But he is happy with the associative sign, which we use to infer what is at present unobservable from what is at present observed:

The associative sign is relied on by living experience, since when a man sees smoke fire is signified, and when he beholds a scar he says there has been a wound. Hence, not only do we not fight against living experience, but we even lend it our support by assenting undogmatically to what it relies on, while opposing the private intentions of the dogmatists [*Ibid.*, II, 102].

The 'associative sign' is that which indicates an object or event which at present is hidden from us, but which we can at least in principle discover later. Sextus has, like a good modern positivist, been opposing the indicative sign, which was supposed to lead us to theoretical entities that are in principle unobservable. His scepticism anticipates many of the concerns of a modern logical empiricist, but we do not find him here enunciating any sceptical problem about the future. Nor can we regard his criticism of inductive generalization as a sceptical problem about induction, for he is opposed only to illegitimate use of the syllogism. This interpretation is supported by Stough's [1969] analysis of these texts.

The sceptical problems about induction arise in quite another context. To understand it, we must retrace some of the ground of Chapters 3–5 above. There are two distinct questions: 'How did probability become possible?' and, 'How did the sceptical problem of induction become possible?' The answer to the first question has primarily to do with a transformation in the mediaeval concept of *opinio*. The result was a concept of 'internal evidence', i.e. of evidence other than testimony. In scholastic epistemology opinion was probable when well attested. Then the world began to testify by its signs. So the probable sign is the sign through which the world gives testimony. Moreover signs may be imperfect and only 'very often' right. Frequency and credibility are thus linked. When conventional and natural sign are finally distinguished, it is the latter that furnish 'internal' evidence. With these transformations in hand, the dual concept of probability was possible. The analytic problem of induction was also possible for as soon as there was a concept of internal evidence, men could start to order the different degrees to which hypotheses are supported. But the sceptical problem of induction remained unknown. To understand why we must examine transformations in the concept of *scientia* or knowledge. Although these are not so essential to the formation of the dual concept of probability, they are integral to the sceptical problem of induction.

Opinion was the staple of low science while knowledge was the goal of high science. Paracelsus was the 'Luther of the physicians', as Copernicus was the Luther of the astronomers. One consequence of their twin revolution was that knowledge and opinion, formerly disparate, entered the same league. Or rather, what happened was that a substantial part of the potential domain of knowledge, including astronomy and the investigation of motion, became part of the domain of opinion. In the writing of Hume, the term 'knowledge' is reserved for pure mathematics. This agrees with the scholastic conception of knowledge as demonstration from first principles. But Aquinas thought one could demonstrate causes and thereby explain why things are as they are. For Hume, demonstration is a matter of the 'comparison of ideas'. This operation can be performed chiefly in the realm of mathematics. Cause, on the other hand, is relegated to the other scholastic category that Hume variably calls 'opinion' or 'probability'. Once the concept of internal evidence was established by 1660, the final transformation needed for the sceptical problem of induction was this transference of causality from knowledge to opinion.

In much modern discussion of Hume it is inadequately noticed how closely, albeit reluctantly, he hews to the established categories of 'knowledge' and 'probability'. A great deal of recent English epistemology has meandered around such questions as 'do I know I am not dreaming now?' or, 'do I know that I have a hand before me?' Philosophers who argue from the meaning of the verb 'know' in ordinary English wish to answer 'yes' to these questions. In the *Treatise* the answer is as a matter of course 'no'. It accepts that what 'knowledge' means is first principles, demonstrations, and comparison of ideas. Hume is certainly attentive to established usage and regrets an inconsistency in it. Probability, from scholastic times, had a pejorative element (as noted e.g. in the quotation from Byrne in Chapter 3). Now that the category of knowledge is relinquishing everything except pure mathematics, the category of 'probability' or of 'opinion' will include items which we cannot complain of as being 'merely probable'. ''Tis however certain, that in common discourse we readily affirm, that many arguments from causation exceed probability, and may be received as a superior kind of evidence.' Modern linguistic philosophers have cited this as the beginning of good commonsense attention to 'common discourse' finally breaking through the clouds of scholasticism. Hume's remark is indicative of something else. Although he employs the categories of knowledge and opinion, he strips the former of causation. In the common discourse of 1739 many arguments derived from cause and effect were not called (merely) probable, because they had been candidates for knowledge which is (*a*) opposed to probability and (*b*) encompasses causes. The latter feature is ended by the time of Hume, but some reasonings from causes retain the former feature. Hence they are not, now, knowledge. Yet they have never been (mere) probability, so 'they may be received as a superior kind of evidence'. Hume calls these 'proofs' that fall short of demonstration, but when he presents his argument about induction he treats them under the head of probability.

It is clear why the sceptical problem of induction requires a transformation in *opinio*: without that, there is no concept of internal evidence about which to be sceptical. It should also be clear why Hume can begin only when causation is stolen from knowledge. So long as causes were the subject of demonstration from first principles, there would still exist necessary connections between cause and effect, and in particular, necessary connections between a present event (a cause) and a future one (an event). The necessary connections were contingently necessary. That is, it is a contingent ｉｉｉｉｉｉｉ ｉ ｉｉｆ ｆ ｉｉ ｉ ｉｉｉｉ ｉｉ ｊｉｉｉｉｉｉｉｉｉｉ ｔｈｅＵＵＦＩ, ＦＵＭＭＭＭ Ａｉｉ ＆ ｇｉｉｉｉｉ ｉｉｄ ｉｉｌ ｌｉｉｌ principles, is in fact the theory of the world. But given that it is the theory, then (to use a modern way of expressing the appropriate notions) the very meanings of the terms in the theory are settled by the theory, and so the propositions of the theory are analytic. That is why I say that in the scholastic view, propositions of cause and effect are contingently necessary. Notoriously Hume spends many pages demolishing the idea of necessary connection. That done, his basic sceptical problem is stated succinctly. An expectation that the future will be like the past must be either knowledge or opinion. But all reasoning concerning the future must be based on cause and effect. Reasoning concerning cause and effect is not knowledge. Therefore it must be opinion, or probability. But all probable reasoning is founded on the supposition that the future will resemble the past, so opinion cannot be justified without circularity. Knowledge and probability are exhaustive alternatives. Hence expectation about the future is unjustified.

To understand the preconditions for this argument we need to investigate knowledge and causation. A proper scrutiny demands a full re-examination of seventeenth century 'high science'. I shall be content with the very end, when the scholastic goals of high science have been severely eroded. Robert Boyle, in making the low science of alchemy into the high science of chemistry had much to do with that erosion. The alchemists, although dreaming of causes, had to be content with signs. They believed that the world worked according to its primary qualities, but they could only experiment on the secondary qualities. There was still the belief that there were true necessary connections among the primary qualities that made everything go. If I may be forgiven the crudeness in such a brief sketch, Boyle, for the first time succeeding in getting behind the phenomena, found no scholastic causes. He speculated about primary qualities, but necessary connections were nowhere in sight. Hence the whole conceptual scheme of a demonstrative knowledge of primary qualities was disintegrating. The final stage in this disintegration gives us an immediate key to a sceptical problem of induction. This is the theory of gravitation. It is only a final stage, and not even an essential one, but it is a good reminder of the state of the old 'knowledge'. In the preceding chapter I have cited numerous Royal Society theologians contending that the non-mechanical law of gravity serves only to describe constant regu-larities in the universe. Newton, the chief glory of physics, has not come up with the goods. Where we had longed for causes and rational demonstration, we found only constant conjunction and lawlike regularity.

Berkeley's reaction is instructive. In a late work, *Siris*, he attacked the corpuscular philosophy of Boyle and Locke on the ground that it never finds efficient causes. It is restricted to seeking 'the general rules and methods of motion and conformity' [1744, p. 111]. Earlier, in *De motu*, written about 1720, he had stated clearly that,

It is not, however, in fact the business of physics or mechanics to establish efficient causes, but only the rules of impulsions or attractions, and, in a word, the laws of motions, and from the established laws to assign the solution, not the efficient cause, of particular phenomena [sec. 35].

This idea of Berkeley's is not fully derived from physics – he is in truth reporting a widespread view that exactly coincides with his philosophy. In Sec. 31 of the *Principles*, published in 1710, he mentions the 'sort of foresight' provided by what are called laws of nature: food nourishes, to sow seed in seedtime is the way to reap the harvest, and so forth. He asserts that all these things we know

182

'not by discovering any *necessary connection* between our ideas, but only by the observation of the *settled laws of nature*'. Moreover, in *Towards a New Theory of Vision*, he indicates that this rejection of necessary connection has far deeper roots than a problem about gravity:

Upon the whole, I think we may fairly conclude that the proper objects of vision constitute an universal language of nature ['an universal language of the Author of nature' in the 3rd edition] whereby we are instructed how to regulate our actions [. . .] It is by their information that we are principally guided in all the transactions and concerns of life. And the manner wherein they signify and mark out unto us the objects which are at a distance is the same with that of languages and signs of human appointment, which do not suggest the things signified by any likeness of identity of nature but only by an habitual connection that experience has made us to observe between them [sec. 147].

'*The connection of ideas does not imply the relation of cause and effect but only of mark or sign with the thing signified. The fire which I see is not the cause of the pain I suffer upon approaching it, but the mark that forewarns me of it.*' With this passage in sec. 65 of the *Principles*, truly, as Michel Foucault says, 'Hume has become possible.'

The knowledge that divined, *at random*, signs that were absolute and older than itself has been replaced by a network of signs built up step by step in accordance with a knowledge of what is probable. Hume has become possible [Foucault 1970, p. 60].

Cause and effect – the paragon of the old knowledge that was demonstration – and signs, the purveyors of opinion, have become one. The sceptical problem of induction is possible. Or rather, in stating the sceptical problem of induction, Hume completed that historical transformation by which the signs of the low sciences became identical with the causes of the high. Berkeley had said that the things we commonly take for causes – such as the fire – are not really causes. They are signs uttered by the Author of the universal language, and that Author is himself the efficient cause. Physics investigates not efficient causes but mere regularities which we know about not by 'any likeness of identity of nature' but by 'an habitual connection'. The causes lie with God. Hume enunciates the final twist. The fire is, after all, the efficient cause, but like all efficient causes it is only a sign!

Hume, then, completes the Berkeleyan syllogism. Causes are signs, but the signs suggest the things signified 'only by an habitual connection'. Reasoning by cause and effect is thereby 'habit and custom only'. We can indeed find such notions verbally prefigured in the coarse philosophizing of those Royal Society theologians who wrote about constant regularity and the new theory of gravitation. But they are not the source of Hume's thinking. They merely express what is happening to the concepts of the time. They conveniently mark the end of the old 'knowledge' because the whole republic of letters begins chanting that the greatest known law of nature is a 'mere constancy' learned by experience which leaves us ignorant of the efficient cause. Yet even in terms of superficial historical 'influence' speculation about gravity did not much move Berkeley. It was an afterthought used for example in *De motu*. Indeed if one examines the main 'influences' on Berkeley's thought one is directed back to the more profound symptoms of the breakdown in knowledge. One of the chief precursors of Berkeley's doctrine, namely Malebranche's theory of occasionalism, was intended to solve the problem of interaction of mind and matter by conceiving of 'feelings' and sensations as signs that God constantly presents to the mind. Although we think of this as philosophical psychology, Leibniz reminds us that it began as physics. After distinguishing minds from matter Descartes had supposed that a mind interacts with material substance at a geometric point, perhaps in the pineal gland. Leibniz insisted that this is bad physics. Descartes had inadequate conservation laws. He knew force is conserved, but did not know that conservation is vectorial. Thus force at a point could, so far as physics was concerned, be directed any way. So the forces which the human body brings into play are determined by the laws of physics, but the direction of their application is extra-physical, that is to say, mental. Only when Leibniz discovered the true conservation laws was he bound to invoke pre-established harmony to replace interaction between mind and matter. Martial Guéroult has amply shown how that doctrine arose chiefly in order to solve problems in dynamics. Long before Hume, and actively rejecting any law of gravity, Leibniz had the idea of 'constant conjunction'. Minds and bodies 'express' each other, and one body, in being, as we say, 'affected' by another, is better described as 'expressing' the other. Arnauld not unnaturally asked Leibniz what this meant. Leibniz replied: 'one thing expresses another, in my use of the term, when there is a *constant and regulated relation* between what is true of the one and what is true of the other'. [*P.S.* ii, 112].

184

Leibniz's philosophy is one of the last desperate defences of the old category of knowledge. He had to believe that there is no interaction between the real things in the universe: there is only 'constant and regular relation'. Moreover, material objects can only be 'well founded phenomena'. He could even write, 'If a thing is not actually sensed, then there is no thing.' Many of the Humeian ideas are present in Leibniz, but one is lacking. For Leibniz, the category of knowledge is still sacrosanct. Truth is ultimately demonstration. Efficient causes may be constant conjunction but final causes will constitute the reason for things. There is a sufficient reason for any truth and it can be proven *a priori*. Where cowards were surrendering the outworks of knowledge to a concept of opinion increasingly fortified by a concept of evidence, Leibniz counter-attacked with one last marvellous innovation. Knowledge had always been demonstration from first principles. Leibniz produced the first 'modern' analysis of proof as formal relationship between sentences. A demonstration of a logically necessary proposition p will be a finite sequence of sentences terminating at p. A proof of a contingent proposition q will be an infinite sequence asymptotically converging to q. Thus all truth is swept into the category of knowledge by refurbishing the concept of demonstration.

Leibniz has been our constant witness to events in probability from 1665 until 1713. He was the first philosopher of probability and anticipated, often in great detail, many of our modern probabilistic conceptions. His lack of anticipation of a sceptical problem about induction – at the very time that he was inventing inductive logic – is as significant a testimony as any. It reminds us that there could be no problem about induction until *scientia* was abandoned. Probability emerged from the Renaissance transformation in *opinio*. That sufficed for an analytic problem about induction. The sceptical problem could arise only when causation had moved from knowledge to opinion. Thus although the emergence of probability is a transformation in opinion, the emergence of 'probability-and-induction' is a more complete event depending on parallel transformations in high science and low science.

BIBLIOGRAPHY

This list provides references for works cited in the text. It also tries to include all published work on probability between 1654 and 1700. The subject has been defined narrowly, excluding the theory of combinations (on which consult Risse [1964]) and the arithmetic mean and associated notions of measurement (on which consult Eisenhart [1971]). It also excludes political economy after Petty, although this is intricately linked to statistical thought. The catalogue to J. M. Keynes' library, donated to King's College, Cambridge, is a good bibliography of that subject.

This list also includes much recent historical work on the emergence of probability, but excludes older histories except when they have become classics, as with Todhunter [1865] and Hendriks [1853–4]. Scholarship written in Russian is also regrettably excluded. For an introduction to that rich source, consult Maistrov [1974] and the works of O. B. Sheynin.

Agricola, Georgius [1556]. *De re metallica.* Basle. (Translated by H. C. and T. H. Hoover, London, 1912.)

Arbuthnot, John [1692]. *Of the Laws of Chance, or, a Method of Calculation of the Hazards of Game*, London. (The author's name does not appear on the title page.)

—— [1710]. An argument for divine providence taken from the constant regularity observed in the births of both sexes. *Philosophical Transactions of the Royal Society of London*, **27**, 186–90.

Archibald, R. C. [1926a]. A rare pamphlet of De Moivre and some of his discoveries, *Isis*, **8**, 671–84.

—— [1926b]. Abraham De Moivre. *Nature*, **117**, 551 (A letter to the editor; see the reply of Karl Pearson 1926).

Arnauld, Antoine [1643]. *La theologie morale des Jesuites.* No place or date on title page. (Two further editions, Paris, 1644. Reprinted in *Oeuvres de Messire Antoine Arnauld*, Paris and Lausanne, 1779, **29**, 74–94).

—— and Nicole, Pierre [1662]. *La logique, ou l'art de penser*, Paris. (For a list of subsequent editions, see the critical edition of the fifth 1683 edition, ed. P. Clair and F. Girbal, Paris, 1965. The fifth edition is translated by P. and J. Dickoff, New York, 1964).

Austin, J. L. [1962]. *Sense and Sensibilia*, Oxford.

Bacon, Francis [1620]. *Instauratio Magna*, London (Pars secunda operis quae dicitur *Novum Organum*, 35–360. There are many translations; the one most often used is by Thomas Fowler, Oxford, 1878, amply reprinted.)

187

Bayes, Thomas [1763]. An essay towards solving a problem in the doctrine of chances, *Philosophical Transactions of the Royal Society of London*, **53**, 370–418. (Facsimile reproduction with commentary by E. C. Molina in *Facsimiles of Two Papers by Bayes*, ed. E. Deming, Washington, D.C., 1940, New York, 1963. Also reprinted with commentary by G. A. Barnard in *Biometrika*, **45**, 293–315, cf. Pearson and Kendall 1970).

Bentley, Richard [1842]. *Correspondence*, ed. C. Wordsworth, London.

Berkeley, George [*Works*]. *The Works of George Berkeley*, ed. A. A. Luce and T. E. Jessop, London, 9 vols, 1948–57. (Includes translations of Latin originals. Page references are to this edition).

—— [1709]. *An Essay towards a New Theory of Vision*, Dublin.

—— [1710]. *A Treatise concerning the Principles of Human Knowledge*, Dublin.

—— [1721]. *De motu sive de motus principio & natura, et de causa communicationis motivum*, London.

—— [1744]. *Siris*. Dublin and London.

Bernard, J. H. [1896]. The predecessors of Bishop Butler, *Hermathena*, **9**, 75–84.

Bernoulli, Jacques (Jakob I or James) [1713]. *Ars conjectandi*, Basle. (There is a complete German translation in Ostwald's *Klassiker* nos. 107–8, and an English translation of Part IV by Bing Sung, Harvard University Department of Statistics Technical Report 2, 1966. Part II is translated in Maseres [1795], while Part I is itself a version of Huygens [1657] which has been translated several times, see below.)

Biermann, Kurt-Reinhard [1954]. Über die Untersuchung einer speziellen Frage der Kombinatorik durch G. W. Leibniz, *Forschungen und Fortschritte*, **28**, 357–61.

—— [1955a]. Über eine Studie von G. W. Leibniz zu Fragen der Wahrscheinlichkeitsrechnung, *Ibid.*, **29**, 110–13.

—— [1955b]. Eine Untersuchung von G. W. Leibniz über die jahrliche Sterblichkeitsrate, *Ibid*, 205–8.

—— [1956]. Speziell Untersuchungen zur Kombinatorik durch G. W. Leibniz, *Ibid*, **30**, 169–72.

—— [1957]. Eine Aufgabe aus den Anfängen der Wahrscheinlichkeitsrechnung, *Centaurus*, **5**, 142–50.

—— [1965a]. Die Behandlung des *Probleme des dès* in den Anfängen der Wahrscheinlichkeitsrechung, *Monatschrifte der Berlinische Akademie der Wissenschaften*, **7**, 70–6.

—— [1965b]. Aus der Entstehung der Fachsprache der Wahrscheinlichkeitsrechnung. *Forschungen und Fortschritte*, **39**, 142.

—— [1967]. Überblick über die Studien von G. W. Leibniz zur Wahrscheinlichkeitsrechnung, *Sudhoffs Archiv*, **51**, 79–85.

—— and Faak, Margot [1957]. G. W. Leibniz *De incerti aestimatione*. *Forschungen und Fortschritte*, **31**, 45–50.

—— [1959]. G. W. Leibniz und die Berechnung der Sterbewahrscheinlichkeit bei J. de Witt, *Ibid.*, **33**, 168–73.

Bibliography

Billettes, M. B. des [1706]. Proposition pour la création des rentes viagères, *Correspondance des controleurs généraux des finances avec les intendants des provinces*, ed. A. M. De Boislisle, Paris, 1883, **2**, 570–8.

Birnbaum, Allan [1967]. John Arbuthnot, *The American Statistician*, **21**, 22–9.

Bloch, Olivier René [1971]. *La philosophie de Gassendi*, The Hague.

Borel, Émile [1909]. *Éléments de la théorie des probabilités*, Paris.

Boudot, P. M. [1967]. Probabilite et logique de l'argumentation selon Jacques Bernoulli, *Les études philosophiques*, N.S. **28**, 265–88.

Boyer, Carl B. [1947]. Note on an early graph of statistical data, Huygens, 1669, *Isis*, **37**, 148–9.

Browne, W. [1714]. *Christiani Hugenii Libellus de Ratiociniis in Ludo Aleae. Or, the value of all chances in games of fortune; cards, dice, wagers, lotteries &c mathematically demonstrated*, London.

Brunet, Georges [1956]. *Le pari de Pascal*, Paris.

Butler, Joseph [1736]. *The Analogy of Religion, Natural and Revealed, to the Constitution and Course of Nature*, London.

Butterfield, Herbert [1957]. *The Origins of Modern Science 1300–1800*, London, 2nd edn. (1st edn. 1949.)

Byrne, Edmund F. [1968]. *Probability and Opinion*, The Hague.

Caramuel, John [1670]. Kybeia, quae combinatoriae genus est, de alea, et ludis fortunae serió disputans, *Mathesis Biceps*, II, Campaniae, 972–1036.

Cardano, Jerome [1663]. *Opera Omnia*, Amsterdam, 10 vols. (Facsimile reprint Stuttgart 1966. Vol. I includes *De ludo aleae*, here published for the first time. Translated by Gould in Ore 1953. This volume also includes *De vita propria*. III includes *De subtiltate*, and IV has *Pactica Arithmetica* and *De proportionibus*. VII includes *De causis signis as locis morborum*).

Cargile, James [1966]. Pascal's wager, *Philosophy*, **41**, 250–7.

Carnap, Rudolf [1950]. *Logical Foundations of Probability*, Chicago.

—— [1952]. *The Continuum of Inductive Methods*, Chicago.

Cheyne, George [1705]. *Philosophical Principles of Natural Religion, containing the Elements of Natural Philosophy and the Proofs for Natural Religion arising from them*, London.

Church, Thomas [1750]. *A Vindication of the Miraculous Powers which subsisted in the Three First Centuries of the Christian Church*, London.

Cohen, L. J. [1970]. *The Implications of Induction*, London.

Collet, François [1848]. *Fait inédit de la vie de Pascal*, Paris.

Condorcet, J. A. N. C. Marquis de [1785]. *Essai sur l'application de l'analyse à la probabilité des décisions rendue à la pluralité des voix*, Paris.

Cotes, Roger [1713]. Editor's preface to the 2nd edn. of I. Newton's *Philosophiae Naturalis Principia Mathematica*, Cambridge. (Reprinted in vol. I of the edition of A. Koyré and I. B. Cohen, Cambridge, 1972, 19–35.)

Cournot, Antoine August [1843]. *Exposition de la théorie des chances et des probabilitiés*, Paris.

Couturat, Louis [1901]. *La logique de Leibniz*, Paris. (Reprinted Hildesheim 1961.)

189

Craig, John [1699a]. *Theologiae Christiane Principia Mathematica*, London. (Chapters 1 and 2 – excluding definitions I–V – are reprinted as Craig's rules of historical evidence, *History and Theory: Studies in the Philosophy of History*, Beiheft **4**, 1964.)

—— [1699b]. A calculation of the credibility of human testimony, *Philosophical Transactions of the Royal Society of London*, **21**, 359–65. (The author's name does not appear.)

Czuber, Emanuel [1899]. Die Entwicklung der Wahrscheinlichkeitstheorie und ihrer Anwendung, *Jahresbericht der deutschen Mathematiker-Vereinigung*, **7**, no. 2.

Darnwell, R. G. [1856]. *A Sketch of the Life and Times of John de Witt, Grandpensionary of Holland, to which is added his treatise on life annuities*, New York.

Davenant, Charles [1699]. *Essay upon the Probable Methods of Making a People Gainers in the Ballance of Trade*, London.

David, Florence Nightingale [1962]. *Games, Gods and Gambling*, London.

Daw, R. H. and Pearson, E. S., [1972], Abraham De Moivre's 1733 derivation of the normal curve: a bibliographical note, *Biometrika*, **59**, 677–80.

Defoe, Daniel [1724]. *The Fortunate Mistress*, London. (Amply reprinted, usually as *Roxana, or The Fortunate Mistress*.)

Deman, T. [1933]. 'Probabilis' au moyen age, *Revue des sciences philosophiques et théologiques*, **22**, 260–90.

—— [1936]. Probabilisme, *Dictionaire de Theologie catholique*, **13**, cols. 417–619.

De Moivre, Abraham [1711]. De mensura sortis, seu, de probabilitate eventuum in ludis a casu fortuito pendentibus, *Philosophical Transactions of the Royal Society*, **27**, 213–64. (Translated in Philosophical Transactions Abridged, **4**, 190–210.)

—— [1718]. *The Doctrine of Chances, or a Method of Calculating the Probability of Events in Play*, London. ('The Second edition, fuller clearer and more correct' London, 1738; reprinted London, 1967. 'The Third edition, fuller clearer and more correct than the former', London, 1756. This also includes the 1743 version of [1725], and is reprinted New York, 1967.)

—— [1725]. *Annuities upon Lives*, London. (There is a 'Second Edition, corrected', London and Dublin, 1730, and a 'Second Edition, plainer fuller and more correct', London, 1743. For further editions and a thorough bibliography of De Moivre, omitting only the 1730 second edition, see Schneider [1968].)

—— [1733]. *Approximatio ad summam terminorum binomii* $(a + b)^n$ *in seriam expansi*, London. (For bibliography see Daw and Pearson [1972].)

De Morgan, Augustus [1838]. On a question in the theory of probabilities. *Transactions of the Cambridge Philosophical Society*, **6**, 423–30.

Dempster, A. P. [1966]. New methods for reasoning towards posterior distributions based on sample data, *Annals of Mathematical Statistics*, **37**, 355–74.

Bibliography

Derham, William [1696]. *The Artificial Clockmaker*, London.

—— [1713]. *Physico-Theology: or a Demonstration of the Being and Attributes of God from his Works of Creation*, London.

—— [1715]. *Astro–Theology: or a Demonstration of the Being and Attributes of God from a Survey of the Heavens*, London.

Diderot, Denis [1746]. *Pensées philosophiques*, The Hague. (The 'Additions, which include Diderot's rude remark LIX on *infini–rien* were written about 1760 and have traditionally been printed at the end of new editions of the *Pensées Philosophiques*, as in *Oeuvres Philosophiques*, ed. P. Vernière, Paris, 1956.)

—— [1765]. *Encyclopédie, ou Dictionnaire raisonné des Sciences, des arts et des métiers*, Paris.

Eisenhart, Churchill [1961]. Boscovich and the combination of observations, *R. J. Boscovitch, Studies of his Life and Work on the 250th Anniversary of his Birth*, ed. L. L. Whyte, London, 200–12.

—— [1963]. The background and evolution of the method of least squares, International Statistical Institute, Ottawa.

—— [1971]. The development of the concept of the best mean of a set of measurements from antiquity to the present day. (Duplicated presidential address to the 131st meeting of the American Statistical Association.)

—— [*Anniversaries*]. Anniversaries in 1965 of interest to statisticians, *The American Statistician*, **19**, Dec., 21–9. (Fermat, Todhunter.) 1966–7, *ibid.*, **21**, April, 32–4; June, 22–9. (Arbuthnot, De Moivre.) 1970, *ibid.*, **24**, Feb., 25–8.

Emerson, William [1776]. *Miscellanies: or, a miscellaneous containing several mathematical subjects*, London.

Fermat, Pierre de [*Oeuvres*] in 4 volumes, ed. by P. Tannery and C. Henry, Paris, 1891–1922. (The 1654 correspondence with Pascal is in Vol. II, 288–331, and includes letters to and from Pascal, Huygens and Carcavi. It is translated by V. Sanford in Smith [1929], 546–65, and by M. Merrington in David [1962], 229–53.)

Fine, T. [1973]. *Theories of Probability*, New York.

Finetti, Bruno de [1937]. La prévision: see lois logiques, ses sources subjectives. *Annales de l'Institut Henri Poincaré*, **7**, 1–68. (Translated in Kyburg and Smokler [1964].)

Foucault, Michel [1970]. *The Order of Things, an Archaeology of the Human Sciences*, London. (Translated from *Les mots et les choses*, Paris, 1966.)

Fracastoro, Giralmo [1546]. *De sympathia et antipathia rerum liber unus: De contagione et contagiosis morbis et eorum curatione, libri iii*. Venice. (Translated by C. Wright as *Fracastoro's De contagione*, New York and London, 1930.)

Freudenthal, Hans [1961]. 250 years of mathematical statistics, *Quantitative Methods in Pharmacology*, ed. H. de Jonge, Amsterdam, xi–xx.

—— [1966]. Aus der Geschichte der Wahrscheinlichkeitstheorie und der mathematische Statistik, *Grundzüge der Mathematik*, ed. H. Behnke et al., Göttingen, vol. IV, 149–95.

Galileo, Galilei [*Opere*]. Edizione Nazionale ed. A. A. Favaro, Florence, 20 vols. in 21, 1890–1909. (*Sopral le scoperte de i dadi* is in VIII, 591–4. It is translated by E. H. Thorne in David, 1962, 192–5.)

—— [1632]. *Dialogo: dove ne i congressi di quattro giornate si discorre sopra i due massimi sistemi del mondo tolemaico, e copernico*, Florence. (Reprint Bari, 1963. Translated by Stillman Drake as *Dialogue Concerning Two World Systems*, Berkeley and Los Angeles, 1953).

—— [1638]. *Discoursi demonstrazione matematiche intorno à due nuove scienze attenenti alla mecanica e i movimenti locali*, Leiden. (Reprint Turin 1958. Translated by H. Crew and A. de Salvio as *Dialogues Concerning Two New Sciences*, New York, 1914, 1952.)

—— [1957]. *Discoveries and Opinions of Galileo*, edited and translated by Stillman Drake, New York.

Gassendi, Pierre [1658]. *Opera Omnia*, Lyon, 6 volumes. (The *Syntagma Philosophicum* are printed, posthumously, for the first time in vols. I and II. There is a facsimile reprint, Stuttgart, 1964. C. B. Brush has translated portions in *The Selected Works of Pierre Gassendi*, New York, 1972. B. Rochot has prepared critical editions with French translation facing, of: *Disquisitio metaphysica*, Paris, 1962, and *Excercitationes paradoxicae adversus aristoteleos*, Paris, 1959.)

Gini, Corrado [1946]. Gedanken zum theorem von Bernoulli, *Schweizerische Zeitschrift fur Volkswirtschaft und Statistik*, **82**, 401–13.

Glanvill, Joseph [1661]. *The Vanity of Dogmatizing: or Confidence in Opinions manifested in a Discourse of the Shortness and Uncertainty of our Knowledge*, London. (Revised as *Scepsis Scientifica*, London, 1665. Both versions reprinted Hove, Sussex, 1970.)

Glass, D. V. [1950]. Graunt's life table, *Journal of the Institute of Actuaries*, **76**, 60–4.

—— [1963]. John Graunt and his 'Natural and Political Observations', *Proceedings of the Royal Society* B, **159**, 2–37. (Reprinted in *Notes and Records of the Royal Society* **19**, 63–100.)

Graetzer, J. [1883]. *E. Halley und C. Neumann*, Breslau.

Graunt, John [1662]. *Natural and Political Observations mentioned in a following Index, and made upon the Bills of Mortality*, London. (Reprinted in *Natural and Political Observations made upon the Bills of Mortality by John Graunt*, ed. W. F. Willcox, Baltimore, 1939. There were further editions in 1663 and 1665. A final, largest edition of 1676 is reprinted in Petty [1899], vol. II. In 1665 there appeared an anonymous tract, pirated from Graunt's book, and titled, *Reflections on the weekly Bills of Mortality for the cities of London and Westminster and the places adjacent: but more especially, so far as it relates to the Plague and other most mortal diseases*, London.)

Gravesande, Wilhelm Jacob's [1774]. *Oeuvres philosophiques et mathématiques*, ed. J. N. S. Allamand, Amsterdam, 2 vols.

Greenwood, Major [1928]. Graunt and Petty. *Journal of the Royal Statistical Society*, **91**, 79–85.

—— [1940]. A statistical mare's nest, *Ibid*; **103**, 246–8.

Bibliography

—— [1941]. Medical statistics from Graunt to Farr, *Biometrika*, **32**, 101–27; [1942] *ibid.*, 203–25; [1943] *ibid.*, **33**, 1–24. (Reprinted in Pearson and Kendall 1970.)

Hass, Karlheinz [1956]. Die mathematische Arbeiten von Johan Hudde (1628–1704), Bürgermeister von Amsterdam. *Centaurus*, **4**, 235–84.

Hacking, Ian [1965]. *Logic of Statistical Inference*, Cambridge.

—— [1971a]. Jacques Bernoulli's '*Art of Conjecturing*', *British Journal for the Philosophy of Science*, **22**, 209–29.

—— [1971b]. Equipossibility theories of probability, *Ibid.*, 339–55.

—— [1971c]. The Leibniz–Carnap program for inductive logic, *Journal of Philosophy*, **68**, 597–610.

—— [1972]. The logic of Pascal's wager, *American Philosophical Quarterly*, **9**, 186–92. (This and the preceding three papers contain, by kind permission of the editors, material used in the present book, chapters 16, 14, 15 and 8 respectively).

—— [1973]. *Leibniz and Descartes: Proof and Eternal Truths*, London, (Reprinted in *Proceedings of the British Academy*, **59**.)

—— [1974]. Combined evidence, *Essays in honour of Stig Kanger*, ed. S. Stenlund, Amsterdam, 113–124.

Hahn, R. [1967]. Laplace's first formulation of scientific determinism in 1773, *Actes du XIe congres international d'histoire des sciences 1965*, **2**, 167–71.

Haldane, J. B. S. [1957]. The Syādvāda system of predication. *Sankhyā*, **18**, 195–200.

Halley, Edmund [1693]. An estimate of the degrees of mortality of mankind, drawn from curious tables of the births and funerals at the city of Breslau; with an attempt to ascertain the price of annuities upon lives, *Philosophical Transactions of the Royal Society*, **17**, 596–610; 654–6.

Hasofer A. M. [1967]. Random mechanisms in Talmudic literature, *Biometrika*, **54**, 316–21. (Reprinted in Pearson and Kendall [1970].)

Heisenberg, Werner [1959]. *Physics and Philosophy*, London.

Hendriks, Frederick [1852–3]. Contributions to the history of insurance, *The Assurance Magazine and Journal of the Institute of Actuaries*, **2**, 121–50; **3**, 93–120.

—— [1863]. Notes on the early history of tontines, *Ibid.*, **10**, 205–19.

Herbert of Cherbury, Edward [1624] *De veritate, prout distinguitur a revelatione, a verisimili, a possibili, et a falso*, London, (Second edition London, 1633. The third edition of 1645 is translated by M. H. Carré, Bristol, 1937).

Hobbes, Thomas [1650]. *Humane Nature, or the fundamental Elements of Policie*, London. (The epistle dedicatory is dated 1640.)

Hudde, Johannes. For correspondence about annuities, see Hendriks [1853–4]; for his annuity table, see Huygens' [*Oeuvres*] VII, 95–6, and letters listed in the entry for Huygens below.

Hume, David [1739]. *A Treatise of Human Nature, being an attempt to introduce the experimental method of reasoning into moral subjects*, London.

—— [1740]. *An Abstract of a Treatise of Human Nature*, London. ('Anonymous'. Reprinted with an introduction by J. M. Keynes and P. Sraffa, Cambridge, 1938.)

—— [1748]. *Philosophical Essays concerning the Human Understanding*, London. (Subsequently titled *An Inquiry concerning the Human Understanding*.)

—— [1779].*Dialogues Concerning Natural Religion*, London.

Huygens, Christian [1657]. *Ratiociniis in aleae ludo*, in *Exercitionum Mathematicorum*, ed. F. van Schooten, Amsterdam. (The Dutch version, which was written earlier, is *Van Rekeningh in Spelen van Geluck*, Amsterdam, 1660. The Latin was reprinted, with an incorrect attribution to Longomontanus, in Caramuel [1670]. It is also the basis of Bernoulli [1713, Book I]. English translations are Arbuthnot [1692], and Browne [1714]. The Dutch version is printed with facing French translation in [*Oeuvres*] XIV.

—— [*Oeuvres*]. *Oeuvres complètes*, The Hague, 22 vols, 1888–1950. (For correspondence about the printing of 1657, see I, nos. 282, 284–5, 288–9, 293, 313; II, 380, 386, 408–9, 411. For 1656 correspondence involving Carcavy, Mylon, Roberval, Fermat and Pascal, see I, nos. 291, 296–7, 301, 306, 308–10, 315, 320, 336, 342. For 1665 correspondence with Hudde on some of the probability exercises, see V, nos. 1374–5, 1384, 1422–3, 1427, 1431, 1434, 1446–50. The 1662 letter from Moray about Graunt [1662] is IV, no. 997; cf. 1013, 1022. The 1669 correspondence with Ludwig about mortality is VI, nos. 1765–6, 1771–2, 1776–8, 1781. The 1671 correspondence with Hudde, arising from de Witt [1671] is VII, nos. 1828, 1839. Hudde's table of mortality is in VII, pp. 96–7.)

James, William [1897]. The will to believe, *The Will to Believe and other Essays in Popular Philosophy*, London and New York.

Jeffner, Anders [1966]. *Butler and Hume on Religion, a comparative analysis*, Stockholm. (Acta Universitatis Upsaliensis, Studia doctrinae christianae Upsaliensia, **7**.)

Jeffreys, Harold [1939]. *The Theory of Probability*, Oxford.

Kellwaye, Simon [1593]. *A defensative against the Plague*, London. (Quoted extensively by W. P. Barrett introducing a facsimile edition of *Present Remedies against the Plague*, London, 1933.)

Kendall, M. G. [1956]. The beginnings of a probability calculus, *Biometrika*, **43**, 1–14. (This and other pieces are reprinted in Pearson and Kendall, [1970].)

—— [1960]. Where shall the history of statistics begin?, *Ibid.*, **47**, 447–9.

Keynes, Geoffrey [1971]. *A Bibliography of Sir William Petty, F.R.S. and of Observations on the Bills of Mortality by John Graunt, F.R.S.*, Oxford. (Lacks Petty [1666].)

Keynes, John Maynard [1921]. *A Treatise on Probability*, London.

Bibliography

King, Gregory [1696]. *Natural and Political Observations and Conclusions upon the State and Conditions of England.* (First published in the 1802 edition of George Chalmers' *Estimate of the Comparative Strength of Great Britain.* Reprinted with an introduction by G. E. Barnett in *Two Tracts by Gregory King,* Baltimore, 1936.)

Kneale, William [1949]. *Probability and Induction,* Oxford.

Knobloch, Eberhard [1971], Zur Herkunft und weiteren Verbreitung des Emblems in der Leibnizschen Dissertatio de arte combinatoria, *Studia Leibnitiana,* **3,** 290–2.

Körner, Stefan [1957] ed. *Observation and Interpretation,* London.

Kohli, Karl [1967]. *Spieldauer, von Jakob Bernoullis Lösung der fünften Aufgabe von Huygens bis zu den Arbeiten von de Moivre,* Inaugural-dissertation (Phil. Fac. II), Universität Zurich.

Koopman, B. O. [1940]. The axioms and algebra of intuitive probability, *Annals of Mathematics,* **41,** 269–92.

Kyburg, H. E. and Smokler, H. E. [1964]. *Studies in Subjective Probability,* New York.

Lambert, Johann Heinrich [*Abhandlungen*] *Logische und philosophische Abhandlungen,* 2 vols, Berlin, 1782–7. (For a study of Lambert's work on probability see Sheynin 1971b.)

—— [1760]. *Photometria sive de mensura et gradibus luminus colorum et umbrae. Augustae Vindelicorum.*

—— [1764]. Phänomenologie oder Lehre von dem Schein, *Neves Organon oder Gedanken über die Erforshung und Bezeichnung des Wahren und dessen Unterscheidung vom Irrthum und Schein,* Leipzig, II, 217–435.

—— [1765]. Theorie der Zuverlässigkeit der Beobachtungen und Versuche, *Beyträge zum Gebrauche der Mathematik und deren Anwendung,* I, 424–88, Berlin. (Zweite verbesserte Auflage, *Beiträge,* Berlin.)

Laplace, Pierre Simon Marquis de [*Oeuvres*]. *Oeuvres complètes,* Paris, 14 vols., 1878–1912. (The *Essaie philosophique sur les probabilités* of 1795 became the Introduction to the *Théorie analytique des probabilités;* it is translated by F. W. Truscott and F. L. Emory, New York, 1902, 1951. The *Essaie* and the *Théorie* form vol. VII of the *Oeuvres* (1886). Laplace's papers on probability commence in 1773 and occur in vols. VIII, IX, XII. WARNING: There is an edition of *Oeuvres* (1843–7), of which the *Théorie* is also volume VII, but with pagination about 10% greater than I have used; *i.e.* 1886 page $n \simeq$ page 1.1(n) of 1847. Todhunter [1865] claims to key his references to the 1820 edition which the 1886 *Oeuvres* vol. VII purports to reprint, but one will often find a reference he cites as page n at about page $n + 5$ in the 1886 vol. VII.)

Lehmann, E. L. [1958]. *Some early instances of Confidence Statements.* Statistical Laboratory, University of California, Berkeley, ONR 5 Technical Report to the Office of Naval Research.

195

Leibniz, Gottfried Wilhelm Freiherr von [*S.S.*]. *Sämtliche Schriften und Briefen*. (This edition, commenced by the Preussischen Akademie der Wissenschaften, and issuing its first volume in 1923, is still in progress. Hence it was often necessary to refer to older editions of Leibniz' work. All of these, with the exception of Klopp, have recently been reprinted by Olms of Hildesheim.)

—— [*Cout.*]. *Opuscules et fragments inédits de Leibniz*, ed. L. Couturat, Paris, 1903.

—— [*Dutens*]. *Opera Omnia*, ed. J. Dutens, Geneva, 1768, 6 vols.

—— [*Erdmann*]. *Opera Philosophica*, ed. J. Erdmann, Berlin, 1839–40.

—— [*Klopp*]. *Werke*, ed. O. Klopp, Hannover, 11 vols, 1864–84. (Vol. V, sec. G contains papers on statistics.)

—— [*M.S.*]. *Leibnizens mathematische Schriften*, ed. C. I. Gerhardt, London, Berlin and Halle, 1849–63, 7 vols.

—— [*N.E.*]. *Nouveaux Essaies sur l'entendement humaine*, in *P.S.* V. (Translated by A. G. Langley, *New Essays concerning Human Understanding*, New York, 1896, 1949.)

—— [*P.S.*]. *Die philosophischen Schriften von G. W. Leibniz*, ed. C. I. Gerhardt, Berlin, 1875–90, 7 vols.

Lindberg, David Charles [1970]. *John Pecham and the Science of Optics*, Madison, Wisconsin.

Locke, John [1690]. *An Essay Concerning Humane Understanding*, London.

Lodge, Thomas [1603]. *A Treatise of the Plague containing the Nature, Signs, and Accidents of the Same*, London.

Lucas, J. R. [1970]. *The Concept of Probability*, Oxford.

Mabbut, George [1686]. *Tables for Renewing and Purchasing of the Leases of Cathedral Churches and Colleges*, Cambridge. (Second edition, corrected, 1700. Many subsequent editions were attributed to Isaac Newton because in the first edition there appears, facing the title page, 'Sept. 10 1685. Methodus hujus Libri rectè se habet, numerique, ut ex quibusdam ad calculum revocatis judico, satis exactè computantur. Is. Newton Math. Prof. Luc.')

Mach, Ernst [1895]. *Popular Scientific Lectures*, Chicago. (Translated by T. J. McCormack from *Populär-wissenschaftliche Vorlesungen*, Leipzig, 1892.)

—— [1960]. *The Science of Mechanics, Critical and Historical Account of its Development*, La Salle, Ill. (Translated by T. J. McCormack from *Die Mechanik in Ihrer Entwicklung*, Leipzig, 1883.)

Madden, E. H. [1957]. Aristotle's treatment of probability and signs, *Philosophy of Science*, **24**, 167–72.

Mahalanobis, P. C. [1957]. The foundation of statistics, *Sānkhyā*, **18**, 183–94.

Mahāvīrācārya [1912]. *The Ganita-sāra-sangraha of Mahāvīrācārya*, with English translation and notes by M. Rangācārya, Madras.

Mahnke, Dietrich [1925]. Leibnizens Synthese von Universalmathematik und Individualmetaphysik, *Jahrbuch für Philosophie und phänomenologische Forschung*, **7**, 305–611.

Bibliography

Maistrov, L. E. [1974]. *Probability Theory*, New York (Translated by S. Kotz from the Russian of 1964.)

Manuel, F. E. [1963]. *Isaac Newton, Historian*, Cambridge.

—— [1968] *A Portrait of Newton*, Cambridge, Mass.

Margolin, Jean Claude [1968]. Science et cosmos dans la pensée de Jerome Cardan, *Actes du XIe Congrès internationale d'histoire des sciences, 1965*, **2**, 253–8.

Maseres, Francis [1795]. *The Doctrine of Permutations and Combinations*, London. (Includes Wallis [1685] and a translation of Part II of J. Bernoulli [1713].)

Mauduit, Michel. [1677]. *Traité de religion contre les athées, les deistes et les nouveaux pyrrhoniens*, Paris. (Only M. M., and not the author's name, appears on the title page.)

Méré, Antoine Gombault [1677]. *Les agrémens*, Paris.

Mesnard, Jean [1965]. *Pascal et les Roannez*, Bruges.

Milman, H. H. [1860]. Translation of *The story of Nala*, ed. by M. Williams, Oxford. (Reissued in the Chowkhamba Sanskrit Studies 53, Varanasi, 1965.)

Mises, R. von [1951]. *Wahrscheinlichkeit, Statistik und Wahrheit*, Berlin. (First edition 1928. Translated by Hilda Geiringer as *Probability, Statistics and Truth*, but page references are to the German edition because the translation renders *gleichmöglich* (Laplace's *également possible*) 'equally likely' which is not suitable for my analysis of equipossibility.)

Montmort, Pierre Rémond de [1708]. *Essay d'analyse sur les jeux de hazard*, Paris ([1714], the second edition, revised and augmented, and including letters between N. Bernoulli, Waldegrave and Montmort, Paris. Reprinted Paris, 1714.)

Newton, Isaac [*Papers*]. *The Mathematical Papers of Isaac Newton*, ed. D. T. Whiteside and M. A. Hoskin, Cambridge, 1967 *et. seq.* (For some notes on geometrical probabilities, see I (1664–6) 58–61.)

—— [*Correspondence*]. *Isaac Newton's Papers and Letters on Natural Philosophy*, ed. I. B. Cohen, Cambridge, Mass., 1958.

—— [1687].*Philosophiae naturalis principia mathematica*, London (2nd edn. Cambridge, 1713, 3rd edn. London, 1726. Facsimile edn. of the 3rd edn. prepared by A. Koyré and I. B. Cohen, Cambridge, 1972.)

—— [1728]. *Chronology of ancient kingdoms amended*, London.

Newton, John [1668]. *The scale of interest*, London.

Neyman, Jerzy [1950]. *First Course in Probability and Statistics*, New York.

Nicholas of Cusa [1967]. *Werke*, ed. P. Wilpert, Berlin, 2 vols. (There is an English translation of *Idiota* as *The Idiot*, London, 1650.)

Obertello, Luca [1964]. *John Locke e Port-Royal, Il problema della probabilità*, Trieste.

Orcibal, M. J. [1956]. Le fragment infini–rien et ses sources. *Blaise Pascal, l'homme et l'oeuvre*, Cahiers de Royaumont, Philosophie **1**, Paris, 159–86.

Ore, Øystein [1953]. *Cardano, the Gambling Scholar*, Princeton (Reprinted New York, no date.)

—— [1960]. Pascal and the invention of probability theory. *American Mathematical Monthly* **67**, 409–19.

Pacioli, Luca [1494]. *Summa de arithmetica, geometria, proportioni et proportionalità*, Venice.

Paracelsus [*Werke*]. *Sämtliche Werke*, ed. K. Sudhoff, 14 vols., Munich, 1922–3. (For some translated selections, see *Paracelsus, Selected Writings*, ed. J. Jacobi, Trans. N. Guterman, London, 1951.)

Pascal, Blaise [1963]. *Oeuvres complètes*, ed. L. Lafuma, Paris (The numeration of the *Pensées* varies a good deal. The critical edition is that of Lafuma, 1951, which is used in the *Oeuvres complétes; infini–rien* is no. 418. Many older editions and translations use Leon Brunschvicg's numeration, with *infini-rien* being no. 233. Several other idiosyncratic numerations are also to be found.)

Pearson, E. S. and Kendall, M. G. [1970]. *Studies in the history of Statistics and Probability*, London.

Pearson, Karl [1924]. Historical note on the origin of the normal curve of errors, *Biometrika*, **16**, 402–4

—— [1925]. James Bernoulli's theorem, *Ibid.*, **17**, 201–10.

—— [1926]. Abraham de Moivre, *Nature*, **117**, 551–2. (A letter of 18 March 1926; cf. Archibald [1926*b*].)

—— [1928]. Biometry and chronology, *Biometrika*, **20(A)**, 241–62, 424.

—— [1929]. Laplace, *Ibid.*, **21**, 202–16.

Petty, William [1666]. Review of Graunt [1662], *Le Journal des Sçavans*, 2 August 1666, 359–70. (There are no pages 363–8. This piece is signed G.P.)

—— [1674]. *The Discourse made before the Royal Society the 26 of November 1674 concerning the Use of Duplicate Proportion*, London.

—— [1683*a*]. *Another Essay in Political Arithmetic, concerning the Growth of the City of London*, London. (Reprinted in 1686 as *An Essay concerning the Multiplication of Mankind together with another Essay in Political Arithmetic*. The stationer's preface says that the implied 'first essay' is lost, but includes a letter from Petty describing its contents. Reprinted in [1699] and [1899] vol. II.)

—— [1683*b*]. *Observations upon the Dublin Bills of Mortality*, London. (Augmented in *Further Observations upon the Dublin Bills*, London 1686, and reprinted [1699] and [1899] vol. II.)

—— [1687*a*]. *Two Essays in Political Arithmetic concerning the People, Housing, Hospitals &c of London and Paris*, London. (A French edition appeared in 1686. Reprinted in [1699] and [1899] vol. II.)

—— [1687*b*]. *Observations upon the Cities of London and Rome*, London. (Reprinted in [1699] and [1899] vol. II.)

—— [1687*c*]. *Five Essays in Political Arithmetic*, London. (In English with French facing. Reprinted in [1699] and [1899] vol. II.)

—— [1699]. *Several Essays in Political Arithmetic*, London.

—— [1899]. *The Economic Writings of Sir William Petty*, ed. C. H. Hull, Cambridge, 2 vols.

—— [1927]. *The Petty Papers*, ed. Marquis of Lansdowne, London, 2 vols. (For a Bibliography of Petty's work, see G. Keynes [1971].)

Bibliography

Peverone, Giobattista-Francesco [1558]. *Due brevi e facili trattati, il primo d'arithmetica, l'altro di geometria*, Lione.

Plackett, R. L. [1958]. The principle of the arithmetic mean, *Biometrika* **45**, 130–5. (Reprinted in Pearson and Kendall [1970].)

—— [1972]. The discovery of the method of least squares, *Ibid*, **59**, 239–51.

Poisson, S.-D. [1837]. *Recherches sur la probabilité des jugements en matière criminelle et en matière civile, precédées des règles générales du calcul des probabilités*, Paris.

Popkin, Richard H. [1953]. Joseph Glanvill: a precursor of David Hume, *Journal of the History of Ideas*, **14**, 292–303.

—— [1960]. *The History of Scepticism from Erasmus to Descartes*, Assen, The Netherlands.

Popper, Karl [1959a]. *The Logic of Scientific Discovery*, London. (First German edition, *Logik der Forschung*, 1935.)

—— [1959b]. The propensity interpretation of probability, *British Journal for the Philosophy of Science*, **10**, 25–42.

Prevost, P. and Lhuillier, S.-A.-J. [1797]. Mémoire sur l'application du calcul des probabilités à la valeur du témoignage, *Mémoires de l'Academie Royale des Sciences et Belles Lettres à Berlin*, 120–51.

Price, Richard [1758]. *A Review of the Principal Questions and Difficulties in Morals*, London. (Second edition, corrected, London, 1769. Third edition enlarged by an appendix, London 1787, reprinted Oxford, 1948. The third edition shows signs of influence by Bayes [1763] which Price ⟨illegible⟩

Ptoukha, M [1937]. John Graunt, fondateur de la démographie, *Congrès international de la population*, Paris, II, 61–74.

Rabinovitch, Nathum L. [1969]. Probability in the Talmud, *Biometrika*, **56**, 437–41.

—— [1970]. Combinations and probability in rabbinic literature, *ibid.*, **57**, 203–5.

Ramsey, Frank Plumpton [1931]. Truth and probability, *The Foundations of Mathematics and other Logical Essays*, ed. R. B. Braithwaite, London. (Written in 1926.)

Randall, John Hermann [1961]. *The School of Padua and the Emergence of Modern Science*, Padua.

Reichenbach, Hans [1935]. *Wahrscheinlichkeitslehre*, Leiden. (Trans. by E. H. Hutten and M. Reichenbach as *The Theory of Probability*, Berkeley, Calif., 1949, 1971.)

Reiersøl, Olav [1968]. Notes on some propositions of Huygens in the calculus of probability, *Nordisk Matematisk Tidskrift*, **16**, 88–91.

Renou, Louis and Filliozat, Jean [1947]. *L'inde classique*, Paris.

Riddell, William Renwick [1928]. *Hieronymous Fracastorious and his Poetical and Prose works on Syphillis, with a full glossary of medical and other terms employed by him*, Toronto.

Risse, Wilhelm [1964 et seq.]. *Die Logik der Neuzeit*, Stuttgart–Bad Cannstatt.

Roberts (or Robartes) Francis [1693]. An arithmetical paradox concerning the chances of lotteries, *Philosophical Transactions of the Royal Society*, **17**, 677–81.

Russell, Bertrand [1948]. *Human Knowledge, its Scope and Limits*, London.

Sambursky, Samuel [1956]. On the possible and probable in ancient Greece, *Osiris*, **12**, 35–48.

Sauveur, Joseph [1679]. Supputation des avantages du banquier dans le jeu de la bassete, *Journal des Sçavans*, pp. 38–45.

Savage, L. J. [1954]. *The Foundations of Statistics*, New York.

Schmitt, Charles B. [1966]. Perennial Philosophy: from Agostino Steuco to Leibniz, *Journal of the History of Ideas*, **27**, 505–32.

—— [1969]. Experience and experiment: a comparison of Zabarella's view with Galileo's in '*De motu*', *Studies in the Renaissance*, **16**, 80–138.

Schneider, Ivo [1968]. Der Mathematiker Abraham de Moivre (1667–1754), *Archive for History of Exact Sciences*, **5**, 177–317.

Seal, H. L. [1949]. The historical development of the use of generating functions in probability theory, *Mitteilungen der Vereinigung schweizerischer Versicherungsmathematiker*, **49**, 209–28.

Sextus Empiricus [*Loeb*]. Greek with facing translation by R. G. Bury, Loeb Classical Library, London and New York, 3 vols, 1933–6.

Shell, E. D. [1960]. S. Pepys, I. Newton and probability, *The American Statistician*, **17**, no. 4, 27–30.

Sheynin, O. B. [1966a]. On selection and adjustment of direct measurements, *Geodesy and Aerophotography* **2**, 114–17. Translated from *Geodeziia i Aerofotos' emka*, 1966, **2**.

—— [1966b]. Origin of the theory of errors. *Nature*, **211**, 1003–4.

—— [1968]. On the early history of the law of large numbers, *Biometrika*, **55**, 459–67. (Reprinted with an afterthought in Pearson and Kendall [1970].)

—— [1970]. Daniel Bernoulli, on the normal law, *Ibid.*, **57**, 199–202.

—— [1971a]. Newton and the classical theory of probability, *Archive for History of Exact Sciences*, **7**, 217–43.

—— [1971b]. J. H. Lambert's work on probability, *Ibid.*, 244–56.

—— [1972a]. D. Bernoulli's work on probability, *Rete*, **1**, 273–300.

—— [1972b]. On the mathematical treatment of observations by L. Euler, *Archive for History of the Exact Sciences*, **9**, 45–56.

—— [1973a]. Finite random sums (A historical essay), *Ibid.*, 275–305.

—— [1973b]. R. J. Boscovich's work on probability. *Ibid.*, 306–24.

—— [1973c]. Mathematical treatment of astronomical observations (A historical essay), *Ibid.*, **11**, 97–126.

Smith, D. E. [1929]. *A Source Book in Mathematics*, New York. (Reprinted New York 1959.)

Spinoza, Benedict de [1882]. *Opera*, ed. J. van Vloten and J. P. N. Land, The Hague, 2 vols. (The *Reekening van Kansen* is in II, 521–4; for a French translation see Huygens *Oeuvres*, XIV, 29–31. The letter to van der Meer is no. 38, II, 145–9. This is translated by A. Wolf in *The Correspondence of Spinoza*, London, 1928.)

Bibliography

Steinhaus, H. [1963]. Probability, credibility, possibility, *Zastosowania Matematyki*, **6**, 341–61.

Stough, Charlotte L. [1969]. *Greek Skepticism*, Berkeley and Los Angeles.

Strowski, Fortunat [1930]. *Pascal et son temps*, Paris, 3 vols. (Fifth edition.)

Struyck, Nicolaas [1716]. *Uytreekening der Kannsen in het Spelen, door de Arithmetica en Algebra, beneevens eene Verhandeling van Looterijen en Interest*, Amsterdam. (For a French translation by J. A. Vollgraf, see *Les Oeuvres de Nicolas Struyck (1687–1769)*, Amsterdam, 1912.)

Sudhoff, Karl [1894]. *Versuch einer Kritik der Echtheit der Paracelsischen Schriften*, vol. I, *Die unter Hohenheims Namen erschienen Druckschriften*, Berlin.

Süssmilch, Johan Peter [1741]. *Die göttliche Ordnung in den Veränderungen des menschlichen Geschlechts aus der Geburt, dem Tode und der Fortpflanzung desselben erweisen*, Berlin. (Enlarged edition, 1761.)

Tartaglia, Nicolo [1556]. *General Trattato di Numeri et Misure*, Venice.

Thatcher, A. R. [1957]. A note on the early solutions of the problem of the duration of play, *Biometrika*, **44**, 515–8. (Reprinted in Pearson and Kendall [1970].)

Thorndike, Lynn [1923–58]. *History of Magic and Experimental Science*, New York, 8 vols.

Todhunter, Isaac [1865]. *A History of the Mathematical Theory of Probability from the Time of Pascal to that of Laplace*, London and Cambridge. (Reprinted New York, 1949.)

Toulmin, Stephen [1950]. Probability, *Proceedings of the Aristotelian Society, Supplementary vol. 24, 27–62. (Reprinted in A Flew, Essays in Conceptual Analysis*, London, 1956.)

Valla, Lorenzo [1922]. *Treatise on the Donation of Constantine*, edited and translated by C. B. Coleman, New Haven, Conn. (This is the Latin text of a 1451 MS., with facing translation. Valla wrote his treatise in 1440.)

Venn, John [1866]. *The Logic of Chance*, London. (Enlarged second edition 1876; further enlarged edition 1886. Reprint of third edition of 1888, New York, 1962.)

Voltaire, Francois Marie Arouet de [1734]. Sur les pensées de M. Pascal. Letter XXV of *Lettres Philosophiques* par M. de V***, Amsterdam. (Also Rouen, 1734. The Basle edition of 1734, equivalent to the London edition in English of 1733, does not include letter XXV. For a reprint and critical study, see *Voltaire: Lettres Philosophiques*, ed. Gustave Lanson, Paris, vol. I, 3rd edn., 1924, vol. II 1917.)

Walker, Helen M. [1934]. Abraham de Moivre. *Scripta Mathematica*, **2**, 316–33. (Also to be found in the reprint of De Moivre [1756].)

—— [1929]. *Studies in the History of Statistical Method*, Baltimore.

Wallis, John [1685]. A discourse of combinations, alternations, and aliquot parts, London. (This is the fourth 'Addition' to *A Treatise of Algebra*, 1685, and is paginated 104–52. Reprinted in Maseres [1795].)

Weiss, H. [1942]. *Kausalität und Zufall in der Philosophie des Aristoteles*, Basle.

Westergaard, Harald L. [1932]. *Contributions to the History of Statistics*, London. (Reprint The Hague, 1969.)

Whewell, William [1837]. *History of the Inductive Sciences*, 3 vols., London. (Third edition, London, 1857, reprinted London, 1967.)

—— [1840]. *The Philosophy of the Inductive Sciences, founded upon their history*, 2 vols., London. (Second edition, London, 1847, reprinted London, 1967.)

White, Colin, and Hardy, Robert J. [1970]. Huygens' Graph of Graunt's data, *Isis*, **61**, 107–8.

Whittle, Peter [1970]. *Probability*, Harmondsworth, Middlesex.

Wilhelm, Kurt [1936]. Chevalier de Méré und sein Verhaltnis zu Blaise Pascal, *Romanische Studien*, **34**, Berlin.

Wilkins, John [1638]. *The Discovery of a World in the Moone, or, a Discourse tending to Prove that 'tis probable there may be another habitable world in that planet*, London.

—— [1641]. *Mercury, or the Secret and Swift Messenger*, London.

—— [1668]. *An Essay Towards A Real Character and a Philosophical Language*, London.

—— [1675]. *Of the Principles and Duties of Natural Religion*, London. (Posthumous publication.)

Wilson, Margaret [1971]. Possibility, propensity and chance: Some doubts about the Hacking thesis, *The Journal of Philosophy*, **68**, 610–17.

Winternitz, M. [1927]. *A History of Indian Literature*, Calcutta. (Translated by S. Ketkar from the German edition of 1908.)

Witt, Jan de [1659]. *Elementa curvarum linearum*, ed. F. van Schooten, Amsterdam.

—— [1671]. *Waerdye van lyf-renten naer proportie van los-renten*, S'Gravenhage. (Reprinted in *Feest-gave van het Wiskundig Genootschap te Amsterdam*, Haarlem, 1879. Most of it is translated in Hendriks [1852–53] and Darnwell [1856] 81–108.)

Wollenschläger, Karl [1932]. Der mathematische Briefwechsel zwischen Johann I. Bernoulli und Abraham de Moivre, *Verhandlungen der Naturforschenden Gesellschaft in Basel*, **43**, 151–317.

INDEX

Dates of birth and death are given for people who figure in the emergence of probability

Index

205